W9-APD-568

Baseball Math

Grandslam Activities and Projects for Grades 4–8

Third Edition

Christopher Jennison

A
GOOD
YEAR
BOOK™

Dedication

To Emily and Keith, my first and best teachers.

Good Year Books are available for most basic
curriculum subjects plus many enrichment areas. For
more Good Year Books, contact your local bookseller
or educational dealer. For a complete catalog with
information about other Good Year Books, please
contact:

Good Year Books
P.O. Box 91858
Tucson, AZ 85752
www.goodyearbooks.com

Illustrations by Doug Klauba.
Book design by Patricia Lenihan-Barbee.
Additional pages designed by Performance Design.
Copyright © 2001, 2005 Christopher Jennison.

Introduction

Welcome to *Baseball Math*, an activity and project book that connects mathematics skills to one of your students' most involving interests. Assignments range from one-page skill reviews to projects designed to involve groups of students for several weeks. Throughout, the emphasis is on how and why—the essence of problem solving—as well as on communication and reasoning. Many activities include Challenge Problems, and numerous open-ended questions enhance critical thinking skills. Although the intended grade-level range of the book is 4 through 8, no levels are assigned to specific activities. Abilities and aptitudes vary widely from grade to grade and from student to student. The Table of Contents includes a listing of the skills developed in each activity and project.

Classroom teachers, as well as researchers, know that most youngsters prefer academic involvements that combine work and play. A study funded by the Alfred P. Sloan Foundation found that sixth-grade students benefited from instruction that linked such interests as art, hobbies, and sports to math and science assignments. This would not have surprised John Dewey, who wrote that "experience is most rewarding when it involves the seemingly contradictory traits of rigor and playfulness."

The lore and numbers of baseball lend themselves perfectly to

contextualized, real-world instruction. In the Morrison Elementary School in Philadelphia, Paula Goldstein posed this problem to her students: "On June 10, 1944, Joe Nuxhall pitched two-thirds of an inning for the Cincinnati Reds. But he had to get permission first from his high school principal. Joe was only fifteen years old, and was the youngest player ever to appear in the major leagues. How many outs did he record that day?"

In the pages that follow, your students will project the value of their baseball card collections, compute averages, compile statistics, design a ballpark, compete in a fantasy baseball league, and much more. Activities in the first section are brief, one-page assignments; the projects in the second section are more extended and well suited to small- and large-group participation. In this latter section, assignments range from estimating players' future performance statistics to a classroom presentation of the poem "Casey at the Bat." Cooperative learning strategies are modeled in many of the projects.

You can use the activities and projects for review, reinforcement, and enrichment. And don't hesitate to assign these activities to the girls in your classroom. The next time you attend a baseball game or watch one on TV, notice how many girls are at the game. Girls are now included in Little League contests, and girls play softball as well as baseball at many levels of competition. One of baseball's many

Introduction

attributes is that it can be played and enjoyed by almost everyone.

Mathematics is a means of investigation, a way of solving problems, and a way of thinking. It is connected to everything. Baseball activities and projects provide a meaningful and motivating context for mathematics instruction. As Dewey proposed, the traits of rigor and playfulness need not be contradictory.

Acknowledgments

Randy Souviney, an old friend, read an early version of this book and offered some very helpful suggestions for improvement. A new friend, Jack Coffland, read a subsequent draft and provided some finishing touches. Publishing friends Bobbie Dempsey, Jenny Bevington, and Tom Nieman inspired and supported me during their tenures at Good Year Books, exemplifying the sort of author-publisher relationship that all authors cherish.

References

Barra, Allen. *Clearing the Bases: The Greatest Baseball Debates of the Last Century.*
New York: St. Martin's Press, 2002.

Baseball Prospectus 2005: Statistics, Analysis, and Insight for the Information Age.
New York: Workman Publishing Co., 2005.

James, Bill. *The New Bill James Historical Baseball Abstract.*
New York: Fireside Books, 2003.

Koppett, Leonard. *Koppett's Concise History of Major League Baseball: Revised and Expanded Edition.*
New York: Carroll & Graf, 2004.

Neft, David, et al. *The Sports Encyclopedia: Baseball 2004.*
New York: Griffin House, 2004.

Ritter, Lawrence S. *The Story of Baseball, 3rd edition.*
New York: HarperCollins, 1999.

Robinson, Ray and Jennison, Christopher. *Greats of the Game.*
New York: Abrams, 2005.

Ward, Geoffrey and Burns, Ken. *Baseball: An Illustrated History.*
New York: Alfred A. Knopf, Inc., 1996.

Contents

From *Baseball Math: Grandslam Activities and Projects for Grades 4–8* published by Good Year Books. Copyright © 2001, 2005 Christopher Jennison.

Contents

Contents

Activities

Place Hitters

From *Baseball Math: Grandslam Activities and Projects for Grades 4–8* published by Good Year Books. Copyright © 2001, 2005 Christopher Jennison.

The Valdez family went to a baseball game in June. Circle the most sensible answers to the questions below. Explain why the other choices are not sensible.

1. What was the cost of a ticket to the game?

    $2.00 $20.00 $80.00

2. How many hours did the game last?

    3   30   300

3. How many baseball caps were sold by one vendor?

    70   700   7000

4. How much did Mr. Valdez pay for a baseball cap?

    $15.00 $35.00 $350.00

5. How many runs were scored in the game?

    9   90   900

6. How many pitchers appeared in the game?

    5   50   500

7. A trumpet player played the National Anthem before the game. How many minutes did the performance last?

    3   30   300

8. How many players sat in the home team's dugout?

    2   20   200

Card Profits

Terry buys and sells baseball cards. She bought a Derek Jeter card for 40¢ and then sold it for 60¢. A month later she bought the card back for 75¢ and then sold it a second time for 90¢.

1. When Terry bought the card for 40¢ and sold it for 60¢, did she make or lose money?

   _____

   How much? _____

2. When Terry bought the card back for 75¢ and sold it for 90¢, did she make or lose money?

   _____

   How much? _____

3. How much money did she make or lose in all?

   _____

4. If Terry bought a card for 30¢, sold it for 50¢, and then bought it back for 70¢, how much would she have to sell it for to make a total of 50¢?

   _____

Circling the Bases

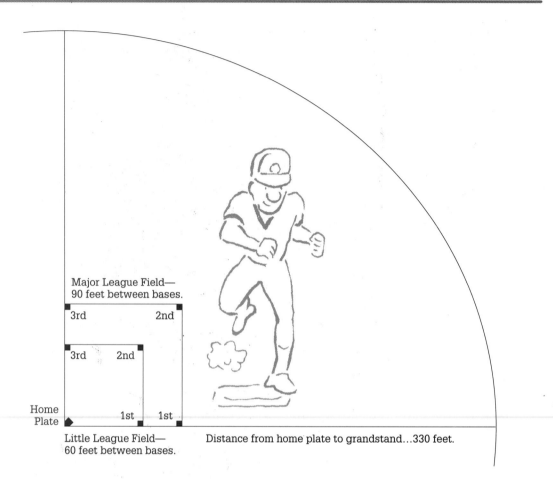

Major League Field—
90 feet between bases.

3rd   2nd

3rd   2nd

Home
Plate   1st   1st

Little League Field—
60 feet between bases.

Distance from home plate to grandstand...330 feet.

Solve the following problems.

1. What is the total number of feet a player would travel around the bases if he hit a home run on a major-league field?

_____

2. What is the total number of feet a player would travel around the bases if she hit a home run on a Little League field?

_____

Challenge problem

3. What is the difference in feet between first base and the grandstand on a major-league field?

_____

What is the difference expressed as a percentage?

_____

From *Baseball Math: Grandslam Activities and Projects for Grades 4–6* published by Good Year Books. Copyright © 2001, 2005 Christopher Jennison.

Change Champs

Look at the prices of the items in the picture. How much change will you receive if you buy the following items with the following amounts of money?

Scorecard $2.50

Popcorn $2.00

Soda $1.75   **T-shirt $17.25**

Baseball $7.50

Hot Dog $2.50

Pennant $4.50

1. Scorecard, change from $3.00.

   _____

2. Pennant, change from $10.00.

   _____

3. Hot dog, change from $5.00.

   _____

4. Soda, change from $2.00.

   _____

5. Hot dog and soda, change from $5.00.

   _____

6. Autographed baseball, change from $10.00.

   _____

7. Popcorn, change from $5.00

   _____

8. Two hot dogs, two sodas, and one popcorn, change from $20.00.

   _____

9. T-shirt, change from $20.00.

   _____

10. Parking fee, at $4.00 an hour: 4 hours parking, change from $20.00.

   _____

Team Trip

Major-league baseball teams regularly travel from one city to another. The table below shows you how far some cities are from each other. Read down and then across to find the number of miles between any two cities.

	St. Louis	Atlanta	Houston	Boston	Chicago
Boston	1180	1084	1961		976
Chicago	296	715	1073	976	
St. Louis		582	780	1180	296
Atlanta	582		875	1084	715
Houston	780	875		1961	1073

1. Chicago and Houston

_____

2. Chicago and Boston

_____

3. St. Louis and Atlanta

_____

4. Houston and Atlanta

_____

5. Boston and St. Louis

_____

6. How far would a team travel if it went from Houston to Atlanta and then on to St. Louis?

_____

7. How far would a team travel if it went from Chicago to St. Louis and then on to Atlanta?

_____

8. Which two cities are closest to each other?

_____

_____

9. Which two cities are the farthest from each other?

_____

_____

10. Which team should allow the most time for traveling?

_____

Why? _____

11. Which team travels the fewest miles in order to play each other team?

_____

From *Baseball Math: Grandslam Activities and Projects for Grades 4–8* published by Good Year Books. Copyright © 2001, 2005 Christopher Jennison.

Diamond Data

Below and on the next page are data files showing what length baseball bat each player should use, according to the player's height and weight.

Boys Baseball Bat Sizes

Batter's Weight in Pounds	Batter's Height							
	3'5"–3'8"	3'9"–4'	4'1"–4'4"	4'5"–4'8"	4'9"–5'	5'1"–5'4"	5'5"–5'8"	5'9"–6'
Under 60	27"	28"	29"	29"				
61–70	27"	28"	29"	29"	30"			
71–80	28"	28"	29"	30"	30"	31"		
81–90	28"	29"	29"	30"	30"	31"	32"	
91–100	28"	29"	30"	30"	31"	31"	32"	
101–110	29"	29"	30"	30"	31"	31"	32"	
111–120	29"	29"	30"	30"	31"	31"	32"	
121–130	29"	30"	30"	30"	31"	32"	32"	
131–140	29"	30"	30"	31"	31"	32"	33"	33"
140–150		30"	30"	31"	31"	32"	33"	34"
151–160		30"	31"	31"	32"	32"	33"	34"
over 160			31"	31"	32"	32"	33"	34"

1. Greg is 5'2" tall and weighs 109 pounds. What size bat should he use?

   _____

2. Jose is 4'7" tall and weighs 82 pounds. What size bat should he use?

   _____

3. Based on your height and weight, what size bat should you use?

   _____

4. In order to use a 32-inch bat, what is the maximum height a player should be?

   _____

5. In order to use a 30-inch bat, what is the minimum weight a player should be?

   _____

6. If you are on a baseball team, check the heights and weights of some of your teammates. Are they using the correct length bat, according to the tables?

   _____

Diamond Data (cont'd.)

Girls Baseball Bat Sizes

Batter's Weight in Pounds	Batter's Height					
	3'10"–4'	4'1"–4'4"	4'5"–4'8"	4'9"–5'	5'1"–5'4"	5'5"–5'8"
Under 40	26"	27"	28"			
40–45	27"	28"	29"	30"		
46–50	27"	28"	29"	30"		
51–60	27"	28"	29"	30"	31"	
61–70	28"	29"	30"	31"	32"	
71–80	28"	29"	30"	31"	32"	33"
81–90	29"	30"	31"	32"	33"	33"
91–100	29"	30"	31"	32"	33"	34"
101–110		30"	31"	32"	33"	34"
111–120		31"	32"	33"	34"	34"
121–130		31"	32"	33"	34"	34"
over 130			32"	33"	34"	34"

From *Baseball Math: Grandslam Activities and Projects for Grades 4–6* published by Good Year Books. Copyright © 2001, 2005 Christopher Jennison.

Write each amount in the questions below with a dollar sign and a decimal point.

1. A baseball batting glove costs four dollars and eighty-nine cents.

   _____

2. A pack of baseball cards costs one dollar and seventy-three cents.

   _____

3. A baseball bat costs twelve dollars and forty-nine cents.

   _____

4. A baseball pennant costs three dollars and nine cents.

   _____

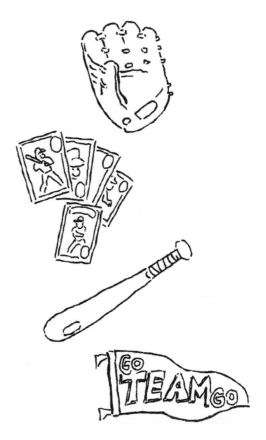

Challenge problems

Find the mystery amount that fits each group of clues below. Use the numbers from the chart.

$1.26	$0.26
$0.62	$1.36
$5.18	$2.95

5. The amount is less than $1.00. The digit in the hundredths place is triple the digit in the tenths place.

   _____

6. The amount is more than $1.00. Four times the digit in the ones place is one less than the digit in the tenths place.

   _____

7. The sum of the tenths and hundredths digits is four more than the ones digit.

   _____

8. There is a six in the hundredths place. The digit in the tenths place is triple the digit in the ones place.

   _____

A Trip to the Park

Solve the following problems.

1. You and your family have decided to go to a baseball game. You must travel from your home in the country to the city where the ballpark is. The distance by highway is 20 miles. If you travel by car at an average speed of 40 miles per hour, how long will it take to get from your home to the ballpark?

   _____

2. The distance from your hometown to the city by train is 25 miles. How long will it take to get to the city if the train travels at an average speed of 50 miles an hour?

   _____

3. Now you can figure out what the difference in cost will be between driving to the ballpark and taking the train. If you travel by car, you will have to pay a toll of $3.00 each way, and $6.00 to park. You must also calculate the cost of gasoline. Remember that the driving distance from your home to the city is 20 miles each way. If your car gets 20 miles to the gallon, and gasoline costs $2.05 per gallon, how much will gasoline cost for the trip?

   _____

4. What will be the total cost of the trip by car?

   _____

5. Now figure out how much it will cost to travel by train. Multiply the number of members in your family by the $6.00 round trip cost of each train ticket. How much will train tickets cost?

   _____

6. Is it cheaper to travel by car or by train?

   _____

7. Think of some other reasons for traveling one way or the other. Is one method of travel safer?

   _____

8. Is one method more relaxing?

   _____

9. Is one method better for the environment?

   _____

From *Baseball Math: Grandslam Activities and Projects for Grades 4–8* published by Good Year Books. Copyright © 2001, 2005 Christopher Jennison.

When he was in high school, the author of this book was a reporter for his town's newspaper, and he reported his high school's baseball games. He was paid 15¢ per column inch for his stories. The calculation was made simply by adding the length of the column or columns that the stories filled.

Times change, and reporters today are paid a lot more than 15¢ an inch for their articles. Imagine that you are a reporter for your town's newspaper, and that you are paid $1.25 per column inch. Attend a high school baseball game and keep notes. Or, if possible, attend a professional game or watch a game on TV.

Write an article about the game. When you have finished, count the number of words in your article. Then look in your local paper and see how many words there are in one inch of a news column. Do this several times. Then find the average number of words in one inch of newspaper-column space. Figure out how many column inches your article would take up.

1. How much money would your article earn?

   _____

2. Suppose you wanted to save $20 a week for a summer trip or to put in a college account. How many 200-word articles would you have to write each week to earn that much?

   _____

3. If you wrote a 500-word article every day for the paper, how much money would you earn each week?

   _____

4. Would you like to be a sportswriter when you get older? Why or why not?

   _____

   _____

   _____

Choosing Sides

Suppose you need names for three Little League teams. The players have suggested six names from which you can choose. One way of allowing more than one selection is by asking players to make first, second, and third choices. Ask seven players to select their first three choices for team names from the names suggested at the right. Next to each team name, write 1 each time a player selects that name as a first choice, 2 when it is the second choice, and 3 when it is the third choice. When you finish, you should have 21 numbers on your chart.

			Student				
Teams	**1**	**2**	**3**	**4**	**5**	**6**	**7**
Braves							
Athletics							
Pirates							
Angels							
Cubs							
Tigers							

Score points for the numbers above as follows:
1 = 5 points; 2 = 3 points; 3 = 1 point
For example, a name that scored 1, 3, 2, 2, and 1 would be worth 17 points.

Write the total scores for each team:

Braves _____

Athletics _____

Pirates _____

Angels _____

Cubs _____

Tigers _____

Which are the three favorite teams?

_____

_____

_____

Why?

_____

_____

_____

_____

From Baseball Math: Grandslam Activities and Projects for Grades 4–8 published by Good Year Books. Copyright © 2001, 2005 Christopher Jennison.

Graphing Favorites

The bar graph shows the favorite teams of the students in Lateisha's class.

Favorite Teams

Dodgers	▬▬▬▬▬▬
Rangers	▬▬
Blue Jays	▬▬▬▬▬▬▬▬▬
Cardinals	▬▬▬▬▬

0 1 2 3 4 5 6 7 8 9 10
Number of Students

1. How many students like the Dodgers best?

   _____

2. How many students like the Rangers best?

   _____

3. How many students like the Blue Jays best?

   _____

4. Which team is liked best?

   _____

5. Which team is liked least?

   _____

6. Do more students like the Dodgers or the Cardinals best?

   _____

Make a bar graph to show the data in the table at the right. Remember to title your graph, list the items the data are about, and decide how to label the number scale. To label the scale, you can count by 2s, 4s, 6s, and so forth.

Number of Baseball Cards Traded

Marty	18
Phyllis	10
Consuela	6
Frank	22
Terry	16
Sonia	12
Kevin	4
Michael	14
Anthony	20
Maria	18
Bart	8

From *Baseball Math: Grandslam Activities and Projects for Grades 4–8* published by Good Year Books. Copyright © 2001, 2005 Christopher Jennison.

Average Attendance

The attendance figures for four baseball teams are shown for three different games.

	Game One	Game Two	Game Three
Robins	9980	3091	11,209
Sparrows	7811	4559	9011
Crows	6744	2985	9165
Ravens	9902	8120	8165

1. What is the Robins' average attendance for the three games?

   _____

2. Find the Ravens' average for the three games.

   _____

3. Find the Crows' average for the three games.

   _____

4. Find the Sparrows' average for the three games.

   _____

5. Find the average for each team's first game.

   _____

6. Find the average for the four teams in the third game.

   _____

7. In a fourth and fifth game, the Sparrows had crowds of 6,642 and 7,198. What was the team's average for five games?

   _____

8. The Crows had 6,719 fans at their fourth game. What was their average attendance for four games?

   _____

9. If the Ravens had a crowd of 10,926 in their fourth game, would their four-game average be more or   less than their average for the first three games?

   _____

   Why?

   _____

   _____

From *Baseball Math: Grandslam Activities and Projects for Grades 4–8* published by Good Year Books. Copyright © 2001, 2005 Christopher Jennison.

Crystal (Base) Ball

From *Baseball Math: Grandslam Activities and Projects for Grades 4–8* published by Good Year Books. Copyright © 2001, 2005 Christopher Jennison.

B onzo the Great, a famous fortune teller, predicts future events. You can help him with some predictions. Solve each problem below by trying and checking combinations of numbers in the chart below. Write the numbers of the combinations after the events. The first one is done for you

Events

4	Player steals eight bases	15	for no money
7	No-hitter pitched	4	on the moon
22	Doubleheader is played	3	by a chimpanzee
5	Players agree to play	11	in one inning

1. This event will happen in 2010. The product of its two numbers is 21, and the sum is 10. What is the event?

   No-hitter pitched by a chimpanzee

2. This event will happen in 2008. The product of its two numbers is 44. The sum is 15. What is the event?

   _____

3. This event will happen in 2021. The sum of its two numbers is 26. The product is 88. What is the event?

   _____

4. This event will happen in 2315. The sum of its two numbers is 20. The product is 75. What is the event?

   _____

Big Bucks

The bar graph below shows how many sports cards were sold between June 2001 and December 2002.

After studying the graph, answer the following questions.

1. Approximately how many cards were sold by the end of 2001?

   _____

2. Approximately how many cards were sold by the end of 2002?

   _____

3. Did the market for sports cards increase or decrease between late 2001 and late 2002?

   _____

4. What was the approximate amount of the increase or decrease?

   _____

5. Do you think the market for sports cards will increase or decrease in the next five years?

   _____

Why?

   _____

   _____

Baseball Card Sales

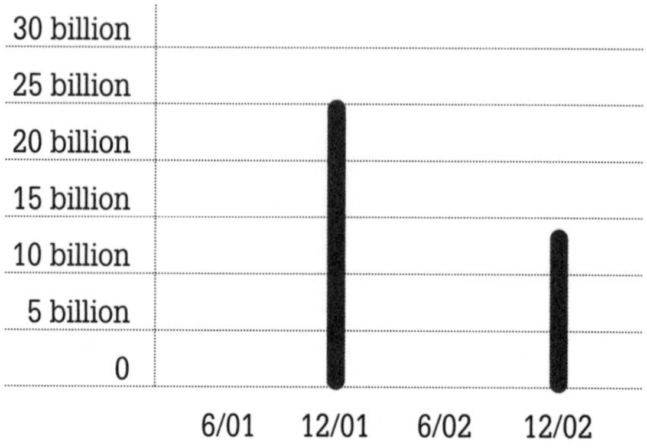

30 billion
25 billion
20 billion
15 billion
10 billion
5 billion
0

6/01   12/01   6/02   12/02

Challenge problem
According to Major-League Baseball properties, three of the companies that issue baseball cards—Upper Deck®, Topps®, and Donruss®—each control about 25% of the baseball-card market. Fleer® has a 15% share and Score® has 10%. In 2002, 14.6 billion baseball cards were sold. How many cards were sold by:

6. Upper Deck®? _____

7. Topps®? _____

8. Donruss®? _____

9. Fleer®? _____

10. Score®? _____

From *Baseball Math: Grandslam Activities and Projects for Grades 4–8* published by Good Year Books. Copyright © 2001, 2005 Christopher Jennison.

Hit Parade

The Tigers, Cardinals, and Indians are keeping track to see which team makes the most hits in a ten-day period. This line graph shows the total number of hits each team made.

1. What was the first day each team was able to make six hits?

   Tigers _____

   Cardinals _____

   Indians _____

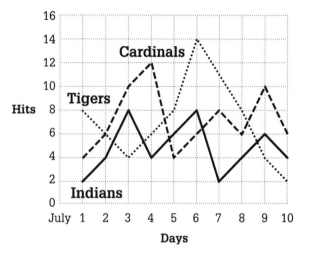

2. On what day did the Tigers and Cardinals have the same number of hits?

   _____

3. On what day were the most number of hits made?

   _____

4. On what day were the fewest number of hits made?

   _____

5. What was the average number of hits made by the Tigers during the ten-game period?

   _____

6. Which team made the most hits during the ten-game period?

   _____

7. Which team improved its hit total the most from one day to the next?

   _____

8. Which team was most consistent throughout the ten-game period? *(In other words, which team had the least variation between its high and low totals?)*

   _____

Time Out

This activity will help you record how many minutes of commercials appear in each half hour of a televised baseball game

1. First, estimate how many minutes of each half hour of a televised ball game are used for commercials. Record your estimate.

2. On the chart below, record the starting and stopping times of each commercial break during the game.

3. Make a bar graph to compare the number of minutes in the game used for commercials and the number of minutes in the entire game. Check to see how close your estimate was.

```
28
26
24
22
20
18
16
14
12
10
 8
 6
 4
 2
 0
```

Game Minutes: _____

Commercial Minutes: _____

Date: _____

Time: _____

Teams: _____

From *Baseball Math: Grandslam Activities and Projects for Grades 4–8* published by Good Year Books. Copyright © 2001, 2005 Christopher Jennison.

Baseball Budgets

An imaginary baseball team budget is shown below. It is divided into percents. Sometimes, instead of using a figure like 10% in a story, a newspaper reporter might say something like: "One of every ten team dollars is spent on maintenance." This is known as a ratio, a relationship, in this case, of 1 dollar to 10 dollars. Keeping that in mind, read the following chart and answer the questions below.

1. What percent of the team's budget is spent on equipment?

   _____

2. Write 6% as a ratio.

   _____

3. One of every five budget dollars goes for

   _____

4. Five out of every hundred budget dollars is spent on

   _____

5. Three dollars of every fifty is spent on

   _____

6. Two out of every twenty-five budget dollars is spent on

   _____

7. One of every three budget dollars is spent on

   _____

8. Write a ratio in words to describe how budget dollars are spent on transportation.

   _____

Expenditures by a Baseball Team

Player salaries	33%
Transportation	20%
Maintenance	10%
Insurance	8%
Equipment	7%
Concessions	8%
Administration	6%
Taxes	5%
Other	3%

Growth Stocks

If you collect baseball cards, chances are you're doing so because you think they'll be valuable in years to come. You could be right. Some cards issued in the last ten years have proven to be quite valuable (although few collectors are lucky enough to acquire cards such as the rookie cards of Nolan Ryan and Reggie Jackson). You can make a rough estimate of what some of your favorite cards and sets will be worth in years to come by using the forms and calculations below.

Begin with the name of the player or set. Then write down the year it was issued. You may not remember what the card or set cost you when you bought it, but you won't be too far off if you use 5¢ for a new card and $50.00 for a set. Below is a sample calculation for Nolan Ryan's rookie card.

Card or set	
Year issued	
Cost new	
Estimated value now	
Years since issued	
Amount of increase	
Percent increase	
Value in one year @20% per year	
Value in five years @20% per year	
Value in ten years @20% per year	

There are several magazines and price guides available that give present values of cards and provide information about the hobby in general.

Remember that the value of a card depends upon its condition and that future values depend on how well a player continues to play and how well the overall market for baseball cards performs. Your estimates will be rough, but it's possible that there are some gems in your collection, so keep your cards in mint condition, and look forward to some pleasant surprises.

Card or set	Nolan Ryan
Year issued	1968
Cost new	$0.05
Estimated value now	$200.00
Years since issued	37
Amount of increase	$199.95
Percent increase	4,000%
Value in one year @20% per year	$240.00
Value in five years @20% per year	$503.00
Value in ten years @20% per year	$1,141.00

Inning Time

Solve the following problems.

1. Estimate the average playing time of one inning of a baseball game.

   _____

2. Now watch or listen to a game and time each inning.

3. Make a bar graph showing the time of each inning.

4. Do you think the first few or last few innings of a game are longer?

   _____

Why do you think so?

   _____

   _____

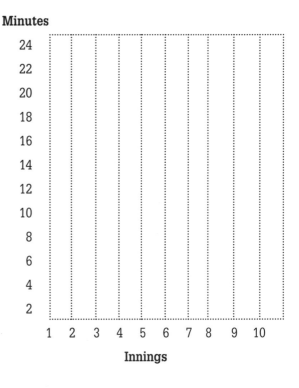

Minutes

24
22
20
18
16
14
12
10
8
6
4
2

1 2 3 4 5 6 7 8 9 10

Innings

Pitching Pros

This activity helps you organize data, or information. The table below shows the names of several big league pitchers and the number of games they won in one season.

First, tally, or write down, each pitcher's number of wins. Then count how many times a number occurred. Write the frequency of each score in the table below. The mode is the total that occurred most often.

1. Which pitcher had the most wins?

   _____

2. How many pitchers won 16 games?

   _____

3. How many pitchers won 18 games?

   _____

4. What is the fewest number of wins?

   _____

5. Which number of wins occurred the least often?

   _____

6. What is the mode?

   _____

Name	Wins
Zito	18
Schmidt	16
Wood	21
Johnson	17
Hudson	18
Martinez	23
Halladay	17
Schilling	15
Zambrano	19
Moyer	19
Mulder	20
Ortiz	16

Tallies	Frequency
14	Two
15	
16	
17	
18	
19	
20	

From Baseball Math: Grandslam Activities and Projects for Grades 4–8 published by Good Year Books. Copyright © 2001, 2005 Christopher Jennison.

Below are the standings in the Galactic League with just a week left in the season:

Team	Won	Lost	Pct.	Games Behind
Cosmos	87	69	.558	
Meteors	85	71	.545	2
Black Holes	82	75	.522	5-1/2
Lasers	81	76	.516	6-1/2

1. The season consists of 164 games. If the Cosmos lose all of the rest of their games, how many games must the Lasers win in order to tie for the lead, assuming the other two teams split their last games?

   _____

2. How many games must the Cosmos win to finish first, no matter what the other teams do?

   _____

3. The Cosmos have eight games left in the season. How many games do the Black Holes have left?

   _____

   If the Cosmos win just three of their last games, and the Black Holes win all of their remaining games, how will the two teams finish, assuming the other teams split their last games?

   _____

Slugfest

Frances was reading an article about a home-run hitting contest. The article gave the following facts about how far each home run went.

Hank hit a homer 3 feet farther than Homer.
Champ hit a homer 3 feet farther than Slugger.
Aaron hit a homer 1 foot less than Hank.
Homer hit a homer 2 feet farther than Slugger.

When Frances turned the page, she was surprised to find that a page was missing. She decided to use the facts above to find out who hit the ball farthest. You can do the same using the table below.

Order of Finish	Names
First (farthest)	
Second	
Third	
Fourth	
Fifth	

From *Baseball Math: Grandslam Activities and Projects for Grades 4–8* published by Good Year Books. Copyright © 2001, 2005 Christopher Jennison.

Game Goodies

You and your sister each have $20 to spend at the ball game. In the chart below, fill in the estimated cost of each item. Then estimate to find the answers to the questions next to the chart. You may want to work with your teacher or classmates.

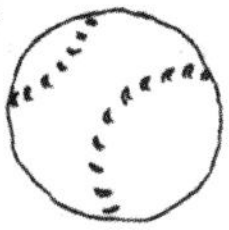

Item	Price
Popcorn	
Ice Cream	
Soda	
Baseball	
Yearbook	
Peanuts	
Hot Dog	
Pennant	
Pretzel	
T-shirt	

Plan how much you will spend on each item. How much do you want to spend on the following?

Food _____

Souvenirs _____

You remember that you will each need $2 to get home. How must you change your plan so that you will each have $2 after the game?

Item	Price
Popcorn	_____
Ice Cream	_____
Soda	_____
Baseball	_____
Yearbook	_____
Peanuts	_____
Hot Dog	_____
Pennant	_____
Pretzel	_____
T-shirt	_____

Lining Up

The pictograph below shows the number of students at Redwood School who are on Little League baseball teams. Each ☺ stands for two students.

Number of Baseball Players

Second Grade	☺
Third Grade	☺ ☺
Fourth Grade	☺ ☺ ☺ ☺ ☺
Fifth Grade	☺ ☺ ☺ ☺ ☺ ☺ ☺ ☺
Sixth Grade	☺ ☺ ☺ ☺ ☺ ☺

1. How many players are in the second grade?

   _____

2. How many players are in the sixth grade?

   _____

3. How many are in the fourth grade?

   _____

4. Which grade has the most players on a team?

   _____

5. Which grade has the fewest players on a team?

   _____

6. How many students at Redwood School are on a team?

   _____

7. Now you can make your own pictograph to show the data in the table below. Remember to:

   - title your graph
   - list the items that the data are about
   - decide on a picture to use, and
   - decide how many votes each picture will stand for

Favorite Team Survey

Team	Number of Votes
Braves	12
Giants	9
Orioles	15
Dodgers	6
Tigers	18
Yankees	21

From *Baseball Math: Grandslam Activities and Projects for Grades 4–9* published by Good Year Books. Copyright © 2001, 2005 Christopher Jennison.

Tips for Tips

You and your family have decided to spend the weekend in the city. You'll be going out to dinner, to a play or movie, and also to a ball game. It is customary to leave a tip for a waiter, a taxi driver, a bellhop, or anyone who provides a service. A standard tip is about 15% of the amount charged by a restaurant or taxi company. To figure a tip for each service below, round off the amount charged for the service to the nearest dollar. For example, in question 1, you would round off the charge, $12.25, to $12.00. In question 3, you would round $7.80 to $8.00.

1. Cost of taxi from airport: $12.25. What should the driver's tip be?

   _____

   What is the total amount you should pay the driver?

   _____

2. Cost of dinner: $63.38. What should the tip for the waiter be?

   _____

   What is the total bill for dinner?

   _____

3. Cost of taxi from hotel to theater: $7.80. What should the driver's tip be?

   _____

   What is the total amount you should pay the driver?

   _____

4. Cost of breakfast: $11.92. Cost of lunch: $27.61. What is the total amount you should pay for both meals?

   _____

5. Cost of taxi from the hotel to the ballpark: $9.60. Cost of taxi from the ballpark to the hotel: $10.35. What is the total amount you should pay for both rides?

   _____

Homer Daze

Every week Tom and Luis keep track of how many home runs are hit in the major leagues during the previous seven days. The bar graph on the right shows the number of home runs hit during one week.

Use the bar graph to answer the questions.

1. What multiples are used for the scale?

   _____

2. On which day were the most home runs hit?

   _____

3. On which day were the fewest home runs hit?

   _____

4. On which day were 11 home runs hit?

   _____

5. How many home runs were hit during the entire week?

   _____

6. What multiples would you use on the scale if you wanted to show the number of base hits made on each day?

   _____

Sunday Attendance Figures

7. Use the blank graph above to make a bar graph that shows the following information:

   Sunday attendance figures
   Braves: 33,000   Yankees: 19,000
   Cubs: 28,500   Expos: 15,000
   Giants: 11,000

From *Baseball Math: Grandslam Activities and Projects for Grades 4–8* published by Good Year Books. Copyright © 2001, 2005 Christopher Jennison.

Collecting Cards

Franklin began a baseball card collection in January. At the end of March, he had 36 cards. At the end of July, he had 84 cards. He collects the same number of cards each month.

1. Make a line graph to show the growth of Franklin's baseball card collection.

2. How many cards did Franklin have at the end of each of these months?

   January _____

   February _____

   March _____

   April _____

   May _____

   June _____

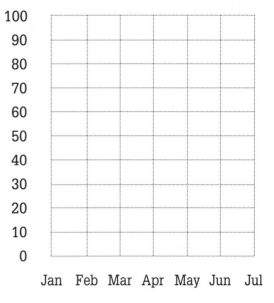

100
90
80
70
60
50
40
30
20
10
0

Jan Feb Mar Apr May Jun Jul

3. At this rate, how many cards will Franklin have at the end of the year?

   _____

4. If Franklin begins with 10 cards and increases his collection each month by 20%, how many cards will he have at the end of twelve months?

   _____

Wait 'til Next Year

Every team wants to improve its record from one year to the next. When a team has a poor season, the players sometimes say, "Wait 'til next year." Below are the 2001 and 2002 won-lost records of some American League teams.

	2001		2002	
	W	L	W	L
New York Yankees	95	65	103	58
Seattle Mariners	116	46	93	69
Minnesota Twins	85	77	94	67
Chicago White Sox	83	79	81	81
Boston Red Sox	82	79	93	69
Detroit Tigers	66	96	55	106
Kansas City Royals	65	97	62	100
Baltimore Orioles	63	98	67	95

1. Which team had the most improved record from 2001 to 2002?

   _____

2. Write that team's improvement as a percentage.

   _____

3. Which team's record had the biggest drop between 2001 to 2002?

   _____

4. What was that team's won-lost percentage in 2001?

   _____

5. What was its won-lost difference in percent from 2001 to 2002?

   _____

   Look up the 2003 and 2004 records for these teams. Which team made the steadiest improvement over the four-year period?

   _____

   Which team's record declined most steadily?

   _____

Home-run Showdown

Solve the following problems.

1. Pretend that during the home-run hitting contest held before the All Star Game, Barry Bonds hit a 440-foot home run, and Carlos Delgado hit a homer 380 feet. How much farther, on a percentage basis, did Bonds's homer go than Delgado's? (Round off your answer to the nearest percentage.)

   23%   63%   16%

2. Pretend also that at that point in the season Bonds's team had played 54% of their games, and Bonds had hit 18 home runs. If Bonds continues to hit home runs at the same pace, approximately how many home runs will he hit by the end of the season?

   44   33   55

3. If Delgado had 531 official at-bats in one season and he hit 35 home runs, what was his home runs to at-bats ratio?

   ❏ One homer for approximately 7 at-bats
   ❏ One homer for approximately 15 at-bats
   ❏ One homer for approximately 4 at-bats

4. If Bonds scored 77 runs, hit 43 home runs, and drove in 109 runs in one season, how many total runs did he produce? (Remember to subtract his homers from his runs and RBI totals.)

   188   207   143   129

5. If Bonds played in 155 games in the season described in number 4, what was the ratio of his runs produced to games played?

   ❏ A little less than one run per game
   ❏ A little more than one run per game

Better Than the Boys?

Solve the following problems.

1. For 11 years beginning in 1943, women played in the All-American Girls Baseball League. The league was made up of teams from around the midwestern part of the country, in cities like Rockford, Illinois; South Bend, Indiana; and Kenosha, Wisconsin. The teams played baseball, not softball, on regulation-sized baseball fields. During the first year, 4 teams played a 108-game schedule. How many games did each team play?

2. Attendance during the first year was almost 108,000 fans. It was estimated that each team drew a larger percentage of their hometown's population than any major league team drew in its best season. In 2004 the New York Yankees' home attendance was 3,775,292 for the year. The population of New York City in 2004 was a bit more than 8,000,000. What percentage of New York City's population attended Yankee games that year?

Challenge problem

3. In 1943 the population of Rockford, Illinois, was about 85,000. How many Rockford fans would have had to attend their girls' baseball games that year in order to have attracted a higher percentage of the city's population than the Yankees did in 2004?

From *Baseball Math: Grandslam Activities and Projects for Grades 4–8* published by Good Year Books. Copyright © 2001, 2005 Christopher Jennison.

The Ballpark Factor

Baseball teams are supposed to play better in their home ballparks. Here are the won-lost records, number of runs scored, and home runs hit by six teams in their home parks during the years shown in the first column.

Years	Team	Won	Lost	Runs	Homers	Opposition Homers
1934–92	Red Sox	2,727	1,898	24,204	4,145	3,687
1976–92	Yankees	789	554	6,191	1,294	1,108
1968–92	A's	1,136	860	8,249	1,677	1,529
1955–92	Braves	1,083	1,059	9,450	2,046	2,011
1916–92	Cubs	3,231	2,786	27,311	4,456	4,272
1962–92	Dodgers	1,437	1,045	9,568	1,605	1,486

1. Which team had the best won-lost percentage in its home ballpark?

_____

2. Which team had the best home run average per season?

_____

3. Which team gave up the most home runs on an annual average basis?

_____

4. In which team's ballpark was it the easiest to hit a home run, based on an annual average of homers hit by the home team and its opponents?

_____

Challenge problems

5. How many times did the team with the best record at home win a league or division championship? *(Look up this information in any baseball record book.)*

_____

6. Do you think the teams with the best overall records have the best won-lost records at home?

_____

Why do you think so?

_____

What else would you need to know in order to answer this question more completely?

_____

Playing Favorites

Which major league baseball player would you predict most people might pick as their favorite?

_____

Now it's time to find out if your prediction is correct. Select eight people to interview about their favorite players.

Name	Age	Favorite Player Choice						

In the "Age" column, fill in one of the following words for each interviewee: Preteen, Teen, or Adult. At the top of each of the first three columns under "Choice," write the name of a player that you think will be mentioned most often. Leave four columns blank for now. Later, during your interviews, use these spaces to write in the names of players you had not previously written.

Conduct your interviews. Compare the predictions you wrote above with the data you collected. Who turned out to be the most popular player? Was any player mentioned by all three age groups? Did more people in your age group agree with your prediction?

From _Baseball Math: Grandslam Activities and Projects for Grades 4–8_ published by Good Year Books. Copyright © 2001, 2005 Christopher Jennison.

Bigger Bucks

Solve the following problems.

1. Starting in 2005, Carlos Beltran, the New York Mets outfielder, will be paid $119 million for seven seasons. What is his average salary per season?

   _____

2. Assuming he plays an average of 130 games each season, what will his salary be per game?

   _____

3. At that rate, Beltran makes more in one game than Mickey Mantle made in his highest-paid season. Do you think Beltran is worth this kind of money?

   _____

4. Why or why not?

   _____

   _____

   _____

   _____

5. Do you think any of the players earning millions of dollars a year are worth the money?

   _____

6. Which ones? Explain why.

   _____

   _____

   _____

   _____

   _____

7. Most players' careers are over by the time they are 35 years old. But they could easily live for another 50 years. Not all players are able to find jobs that pay well after they retire from baseball. Assume that a former player needs $75,000 a year to support himself and his family. How much money will he need to support himself and his family at that level for 50 years, disregarding inflation and interest earnings?

   _____

8. If the player only played in the major leagues for ten years, what would his average salary have to be in order to support himself and his family after he retires at $75,000 a year for 50 years?

   _____

Skill Sharpeners

Six members of the Dolphins Little League team kept track of how many hours they practiced in one week. The chart below shows the number of hours each player practiced.

Student	Batting	Fielding	Bunting	Throwing	Pitching
Todd	5-1/2	3-1/6	2-1/3	0	1
Diana	1-1/6	1-1/6	1-1/2	3-2/3	0
Josh	2-1/2	1-2/3	1-3/4	3	0
Ruben	1-1/6	1	3/4	5-2/3	1
Benji	2-1/2	1-2/3	1	2-5/6	0
Connie	4-3/4	3-1/6	1-3/4	0	1

Choose an operation to solve each of the problems below. Then solve the problem.

1. How many hours did Todd spend batting and fielding?

   _____

2. How much more time did Connie spend fielding than Josh?

   _____

3. How much less time did Ruben spend bunting than Diana?

   _____

4. If Benji spent the same amount of time fielding every week, how many total hours would he spend fielding after four weeks?

   _____

5. How much more time did Ruben spend throwing than Benji?

   _____

6. If Diana spent the same amount of time fielding and throwing each week, how many total hours would she spend on both after three weeks?

   _____

Uniform Numbers

Kelsey, Joe, Sonia, and Greg each have a favorite baseball uniform number. The numbers are 7, 11, 50, and 42. Use the clues and the chart below to find each person's favorite number.

	Kelsey	Joe	Sonia	Greg
7				
11				
50				
42				

Clue 1. Joe's number is a multiple of 10. His number must be 50. Write YES next to 50 under Joe's name. Then write NO in the three other boxes under Joe's name. Write NO next to 50 under Kelsey, Sonia, and Greg.

Clue 2. Both Kelsey's and Sonia's favorite numbers are odd numbers. Only 7 and 11 are odd, so Greg's favorite must be 42.

Clue 3. Sonia's number is larger than Kelsey's. What is Kelsey's favorite number?

_____

What is Sonia's favorite number?

_____

Complete the chart below to find each person's favorite number. Use the clues below.

Clue 1. Gary's number is odd.

Clue 2. Lionel's number is not a multiple of 5.

Clue 3. Chris's number is four times Ellen's number.

	Ellen	Chris	Gary	Lionel
4				
17				
25				
100				

Long Arms

Look at the table below. In the "Best" column, fill in the longest distance each student threw the ball.

Student	1st throw	2nd throw	3rd throw	Best
Carrie	25.73m	26.35m	26.4m	
Roy	15.7m	15.47m	15.28m	
Kyle	20.2m	21.18m	18.2m	
Ingrid	15.07m	26.34m	26.24m	
Eduardo	23.7m	26.34m	26.24m	

1. Whose throws were the most consistent in length (who showed the least variation between throws)?

   _____

2. Which student showed the most improvement from his or her first throw to his or her last throw?

   _____

3. The school record for the baseball throwing contest is 39.68 m. How much farther would Carrie have to throw the ball to tie the record?

   _____

4. What is the difference between Roy's best throw and Eduardo's shortest throw?

   _____

5. What is the total of Kyle's best throw and Roy's shortest throw?

   _____

6. Whose throws are the least consistent (who showed the most variation between throws)?

   _____

Below are some of Carlos Beltran's statistics for 2004:

At bats 599
Hits 160
Runs 121
Homeruns 38
Strikeouts 101

1. Beltran played in 159 games in 2004. His salary in 2005 is $17,000,000. Based on that salary figure, and 159 games played, how much would Beltran earn for every hit?

   _____

2. For every strikeout?

   _____

3. You can look up players' salaries on the Internet. See how much your favorite player earned in 2004, and calculate how much he earned per hit, homerun, and run-batted-in.

   _____

4. In 2004 Ichiro Suzuki set an all-time record for hits with 262. His salary that season was $6,500,000. How much did he earn for each hit?

   _____

5. Based on that figure, do you think Suzuki gives a better value to his team than Beltran?

   _____

6. Why or why not?

   _____

   _____

Challenge Problem
See if you can name a player that delivers a better value to his team than Beltran. Look at his salary for 2004, and calculate how much he earned per hit, homerun, and run-batted-in. Compare those statistics with Beltran's for 2004. Which player do you think delivers a better value, and why?

_____

_____

_____

From *Baseball Math: Grandslam Activities and Projects for Grades 4–8* published by Good Year Books. Copyright © 2001, 2005 Christopher Jennison.

Striking Out

A good pitcher will strike out twice as many batters as he or she walks for a ratio of 2.00 to 1.00. The table below shows the strikeout and walk records of 6 pitchers.

Player	Strikeouts	Walks	Innings
Paul	46	32	109
Carlos	29	18	67
Patty	31	19	88
Hector	33	25	71
Phil	19	11	38
Karen	27	24	52

1. Who has the best strikeout-to-walk ratio?

   _____

   What is that ratio, rounded off to two places?

   _____

2. Who has the lowest strikeout-to-walk ratio?

   _____

   What is that ratio, rounded off to two places?

   _____

3. Patty pitched a total of 88 innings. What was her average number of strikeouts per inning?

   _____

4. Based on this average, how many strikeouts would she have after 134 innings?

   _____

5. Paul pitched a total of 109 innings. What was his average number of walks per inning?

   _____

6. Based on this average, how many walks would he have given up after 202 innings?

   _____

7. Which pitcher had the best walks-per-inning ratio?

   _____

   What is it, rounded off to two places?

   _____

8. Which pitcher would you like to have on your team?

   _____

   Why?

   _____

From *Baseball Math: Grandslam Activities and Projects for Grades 4–8* published by Good Year Books. Copyright © 2001, 2005 Christopher Jennison.

Solve the following problems.

1. In Lions Stadium there are 8 sections. Each is represented by a letter. If 1,537 people are sitting in section A and there are 53 rows in that section, how many people can sit in each row?

   _____

2. If 3,060 people are sitting in Section H and there are 51 rows in that section, how many people can sit in each row?

   _____

3. If there are 3,672 people sitting in Section C and that section has 12 rows, how many people can sit in each row?

   _____

4. Section D has 24 rows. If a total of 2,520 people are sitting in that section, how many people can sit in each row?

   _____

5. If there are 2,160 people sitting in Section G, and 72 people are sitting in each row, how many rows are in that section?

   _____

6. If there are 840 people sitting in Section E and 42 people are sitting in each row, how many rows are in that section?

   _____

7. A total of 5,488 people were seated in Section B on Sunday. That section normally holds 6,104 people in its 56 rows when it's full. How many people can sit in each row?

   _____

8. A total of 4,740 people were seated in Section E on Sunday. That section holds 5,530 people in its 79 rows when it's full. How many people can usually sit in each row?

   _____

Ballpark Field Trip

For a change, you will have a chance to write a question after you have been given the answer.

1. 564 students from Wilson School are going to the ball game on buses. Each bus holds 48 students. The answer is 12. What is the question?

   _____

   _____

   _____

2. Now the answer is the remainder, 36. What is the question?

   _____

   _____

   _____

3. The ballpark has 4,500 seats. Each row holds 200 seats, except for the last row. Write a question so that the answer is the remainder, 100.

   _____

   _____

   _____

4. Write a question so that the answer is 22.

   _____

   _____

   _____

5. Write a question so that the answer is 23.

   _____

   _____

   _____

From *Baseball Math: Grandslam Activities and Projects for Grades 4–8* published by Good Year Books. Copyright © 2001, 2005 Christopher Jennison.

Finding the Average

From *Baseball Math: Grandslam Activities and Projects for Grades 4–9* published by Good Year Books. Copyright © 2001, 2005 Christopher Jennison.

The Dynos baseball team played 20 games last season.

1. Find the team's average score for each five-game stretch. Use paper and pencil or a calculator.

Five-Game Stretch	Scores	Average
First five games	3, 11, 9, 4, 3	
Second five games	2, 19, 3, 7, 9	
Third five games	6, 12, 1, 5, 1	
Fourth five games	4, 13, 4, 9, 5	

2. Find the average number of hits for each player in the table below.

Player	Hits	Average
Ronnie	4, 2, 5, 3, 1	
Jenny	2, 3, 1, 1, 3	
Tom	5, 1, 1, 4, 4	
Bobbie	2, 3, 3, 4, 3	
Julio	4, 4, 5, 2, 5	

3. Find the average attendance for the five games in the table below.

Game 1	12,887
Game 2	11,040
Game 3	9211
Game 4	4530
Game 5	4982
Total	
Average	

4. Find the average number of strikeouts for each player in the table below.

Player	Strikeouts	Average
Len	11, 9, 6, 12, 7	
Marian	7, 9, 9, 10, 5	
Sammy	10, 5, 8, 15, 7	
Chuck	12, 4, 7, 11, 6	
Stan	6, 3, 8, 4, 4	
Holly	8, 5, 9, 10, 3	
Michael	6, 8, 4, 9, 3	

Time for Baseball

Y ou will need a friend to complete this activity with you. Below is a map of the United States divided into four time zones.

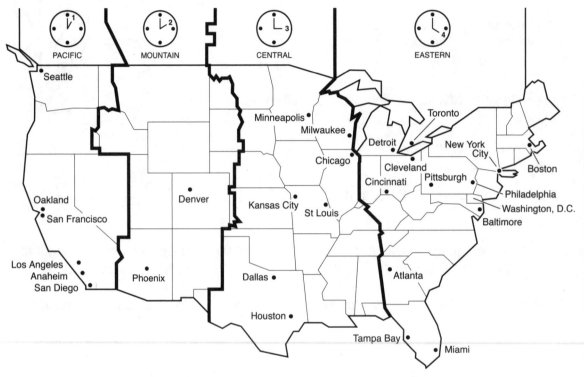

1. Write the names of the 28 cities on cards or small pieces of paper. Then, on 14 other cards or pieces of paper, write an hourly time, such as 3 p.m. Write a different hour on each card.

2. Mix each set of cards or sheets separately, and place them facedown in two stacks. Pick up a city card and a time card (for example, a card for Chicago and a card for 6 p.m.).

3. Ask a friend to pick up a city card. Based on the time it is in your city (6 p.m.), what time is it in your friend's city?

_____

4. Reverse the process, with your friend selecting both a city and a time card and you choosing only a city card. What time is it in your city based on the time in your friend's city?

_____

5. Keep track of how many correct answers each of you makes. This is an activity you can play many times, because the number of combinations of cities and times is high.

Red Hots

Pretend that you are in charge of the concessions sold at your favorite team's ballpark. Concessions include everything from hot dogs to souvenir hats. You'll need to decide how much to charge for the items based on how much they cost you. For instance, suppose hot dogs cost you 20¢ each, and in order to cover your operating costs, you need to increase or "mark up" each hot dog by 200%. That means the cost to the customer must be 200% higher than what you paid for each hot dog.

1. What will you have to charge?

   _____

2. Now calculate the selling prices for the following items based on their costs to you and the percentage profit you must make in order to cover your costs of doing business.

	Cost	Markup	Price
Scorecard	$.25	75%	
Hats	$2.00	50%	
Autographed Ball	$3.50	40%	

Challenge problem

3. For a little more figuring, provide the missing numbers for the items below.

	Cost	Markup	Price
Soda Pop	$.30		$.90
Ice Cream		50%	$1.50

Low Points

It's your sad duty to write a report about the worst hitters and pitchers in history. Their records appear below. To make the project a bit more interesting, some of the numbers are missing, and you'll need to fill them in, based on clues already given.

Name	At Bats	Hits	Ratio	Average	Ranking
Butter Fingers	1962	287	287/1962		.146
O. U. Stink	2885	191	191/2885		
Shakee Batts	786		63/786		
Lowzee Hidder	6889		290/		
U. R. Badd		238	/4117		
Billy Goat	3998	328			
Hesa Whiffer			101/6632		
Bench Warmer	766	22			

	Won	Lost	Ratio	W–L Pct.	Ranking
Lefty Failure	19	188	19/188	.101	
Knuckles Flimsy	77	789			
Outa Control	32	411			
Bull Pen	66	972			
I. M. Crummy			22/619		
Uneeda Curve		452	71/		

Now write a story that includes the ranking order of these players and some information as to why they were able to last in baseball for so long. (You'll have to stretch your imagination for this part.) Then write about what one of the players did after he retired from the game. If you want, you could make up an interview with one of these players. The questions and answers could be quite creative!

From *Baseball Math: Grandslam Activities and Projects for Grades 4–8* published by Good Year Books. Copyright © 2001, 2005 Christopher Jennison.

Ruth on the Mound

Everybody knows that Babe Ruth was a great slugger. And most people know that he was a good outfielder too. But not everyone knows that he was an outstanding pitcher at the beginning of his career, when he played for the Boston Red Sox.

Year	Team	Won	Lost	Pct.	G	GS	CG	SO	SH	ERA
1914	Red Sox	2	1	.667	4	3	1	3	0	3.91
1915	Red Sox	18	8	.692	32	28	16	112	1	2.44
1916	Red Sox	23	12	.657	44	41	23	170	9	1.75
1917	Red Sox	24	13	.649	41	38	35	128	6	2.01
1918	Red Sox	13	7	.650	20	19	18	40	1	2.22
1919	Red Sox	9	5	.643	17	15	12	30	0	2.97

G=Games   GS=Games Started   CG=Complete Games   SO=Strikeouts   SH=Shutouts   ERA=Earned Run Average

1. Above is the Babe's record as a pitcher. Using the years 1914 to 1917, try to calculate what his record would have been if he had been a pitcher for his whole career.

   _____

2. Ruth played a total of 22 years in the majors. Calculate an average of the pitching statistics Babe recorded between 1915 and 1918. Then you can estimate what he might have achieved over a 22-year career. How many games would he have won? (Remember that this is just an estimate; there is no right answer.)

   _____

3. Where would this have put him on the all-time list of pitching victory leaders? (Look up the list in any baseball record book.)

   _____

4. What would his career earned run average have been?

   _____

5. How many strikeouts would he have had?

   _____

6. Where would this have put him on the all-time list of strikeout leaders?

   _____

7. What would have been his ratio of complete games to games started?

   _____

Game Plan

Below is the Boston Red Sox 2004 schedule. This is just an illustration to help you create a schedule for an imaginary league of four teams for a two-month period. Each team must play the other teams 18 times, 9 times at home and 9 times on the road. Create your own schedule using the forms on the next page. You should indicate home games by adding a color to the date square. Begin with one team and create its schedule with the other three teams. Depending on which months you use, each team will have six or seven days off, so space them through the schedule. You can also indicate day games and night games. Most teams play night games during the week and day games over the weekend, but you can arrange day or night games on any days you want.

APRIL

SUN	MON	TUE	WED	THU	FRI	SAT
				1	2	3
N 4 BAL 8:05	5	N 6 BAL 3:05	N 7 BAL 7:05	N 8 BAL 3:05	U 9 TOR 7:05	N 10 TOR 7:05
N 11 TOR 2:05	12	N 13 BAL 7:05	N 14 BAL 7:05	N 15 BAL 7:05	F 16 NYY 8:05	N 17 NYY 1:20
N 18 NYY 2:05	N 19 NYY 11:05	N 20 TOR 7:05	N 21 TOR 7:05	N 22 TOR 7:05	U 23 NYY 7:05	N 24 NYY 1:05
N 25 NYY 1:05	26	N 27 TAM 7:05	N 28 TAM 7:05	N 29 TAM 7:05	U 30 TEX 8:05	

MAY

SUN	MON	TUE	WED	THU	FRI	SAT
					N 1 TEX 8:05	
E 2 TEX 8:05	N 3 CLE 7:05	N 4 CLE 7:05	N 5 CLE 7:05	N 6 CLE 7:05	U 7 KC 7:05	N 8 KC 1:20
N 9 KC 2:05	N 10 CLE 7:05	N 11 CLE 7:05	N 12 CLE 7:05	N 13 TOR 7:05	N 14 TOR 7:05	N 15 TOR 1:05
N 16 TOR 1:05	17	N 18 TAM 7:15	N 19 TAM 7:15	N 20 TAM 7:15	U 21 TOR 7:05	N 22 TOR 7:05
TOR 23 2:05 N	24	N 25 OAK 7:05	N 26 OAK 7:05	N 27 OAK 7:05	U 28 SEA 7:05	F 29 SEA 1:20
SEA 30 2:05 N	31					

JUNE

SUN	MON	TUE	WED	THU	FRI	SAT
		N 1 ANA 10:05	N 2 ANA 10:05	3	U 4 KC 8:10	N 5 KC 7:10
N 6	7	N 8 SD 7:05	N 9 SD 7:05	N 10 SD 7:05	U 11 LA 7:05	F 12 LA 3:15
E 13 LA 8:05	14	N 15 COL 9:05	N 16 COL 9:05	N 17 COL 3:05	U 18 SF 10:15	F 19 SF 3:20
N 20 SF 4:05	21	N 22 MIN 7:05	N 23 MIN 7:05	N 24 MIN 1:05	U 25 PHI 7:05	F 26 PHI 1:20
N 27 PHI 2:05	28	N 29 NYY 7:05	N 30 NYY 7:05			

JULY

SUN	MON	TUE	WED	THU	FRI	SAT
				N 1 NYY 7:05	U 2 ATL 7:35	N 3 ATL 7:05
W 4 ATL 1:05	5	N 6 OAK 7:05	N 7 OAK 7:05	N 8 OAK 7:05	U 9 TEX 7:05	N 10 TEX 7:05
N 11 TEX 2:05	12 ALL-STAR BREAK (Houston)	13	N 14 ANA 10:05	N 15 ANA 10:05	N 16 ANA 10:05	N 17 ANA 10:05
N 18 ANA 4:05	N 19 SEA 10:05	N 20 SEA 4:35	N 21 BAL 7:05	22	W 23 BAL 1:05	F 24 NYY 7:05
N 25 NYY 2:05*	N 26 BAL 7:05	N 27 BAL 7:05	N 28 BAL 7:05	29	U 30 MIN 8:10	N 31 MIN 7:10

AUGUST

SUN	MON	TUE	WED	THU	FRI	SAT
W 1 MIN 2:10	N 2 TAM 7:15	N 3 TAM 7:15	N 4 TAM 7:15	5	U 6 DET 7:05	N 7 DET 7:05
N 8 DET 1:05	N 9 TAM 7:05	N 10 TAM 7:05	N 11 TAM 1:05	N 12 CHW 7:05	U 13 CHW 7:05	N 14 CHW 7:05
N 15 CHW 2:05*	N 16 TOR 7:05	N 17 TOR 7:05	N 18	19	U 20 CHW 8:05	N 21 CHW 1:20
N 22 CHW 7:05	N 23 TOR 7:05	N 24 TOR 7:05	N 25 TOR 7:05	N 26 DET 7:05	W 27 DET 7:05	N 28 DET 7:05
N 29 DET 2:05	30	N 31 ANA 7:05				

SEPTEMBER/OCTOBER

SUN	MON	TUE	WED	THU	FRI	SAT
		N 1 ANA 7:05	N 2 ANA 7:05	U 3 TEX 7:05	N 4 TEX 1:20	
N 5 TEX 2:05	N 6 OAK 10:05	N 7 OAK 10:05	N 8 OAK 10:05	N 9 SEA 10:05	U 10 SEA 10:05	N 11 SEA 10:05
N 12 SEA 4:05	13	N 14 TAM 7:05	N 15 TAM 7:05	N 16 TAM 7:05	U 17 NYY 7:05	F 18 NYY 1:20
N 19 NYY TBA	N 20 BAL 7:05	N 21 BAL 7:05	N 22 BAL 7:05	N 23 BAL 7:05	U 24 NYY 7:05	F 25 NYY TBA
NYY 26 2:05 N / BAL 3 1:35 N	N 27 TAM 7:15	N 28 TAM 7:15	N 29 TAM 7:15	30	U 1 BAL 7:35	F 2 BAL 4:35

Boston Red Sox Baseball Schedule. Reprinted by permission of the Boston Red Sox.

Sun	Mon	Tue	Wed	Thu	Fri	Sat

Month:

Sun	Mon	Tue	Wed	Thu	Fri	Sat

Month:

Sun	Mon	Tue	Wed	Thu	Fri	Sat

Month:

Sun	Mon	Tue	Wed	Thu	Fri	Sat

Month:

Sun	Mon	Tue	Wed	Thu	Fri	Sat

Month:

Antique Cards

Collecting cards is hardly a recent phenomenon, but the hobby was not always pursued by young people. Some of the earliest cards were included in packs of cigarettes. In later years cards were put in gum packs and Cracker Jack boxes. Cards have also been distributed in vending machines, cereal boxes, loaves of bread, and boxes containing baseball equipment, and have even been given away by police officers.

Cartoons were popular, as were "triple folders," in which two player pictures framed an action photo. The front of the card showed a player in full length. About a third of the way down, the card was scored for easy folding. When the top one-third of the card was folded forward and down, another player was shown whose knees connected with the lower legs of the original player. Brief statistics were provided for each player on the back two-thirds of the card.

Perhaps you could research the career records of one or two old players, or write a theme on how old cards compare with modern ones. You might compare format and distribution approaches, as well as gimmicks.

From *Baseball Math: Grandslam Activities and Projects for Grades 4–8* published by Good Year Books. Copyright © 2001, 2005 Christopher Jennison.

The baseball stadiums shown here span more than one hundred years, from the early 1890s to 2005. The ballpark below is the Polo Grounds, once located in upper Manhattan in New York City. The park was built in the 1890s and was remodeled and expanded several times during the early 1900s. It was torn down in 1964 to make room for apartment houses.

Opposite the Polo Grounds diagram is a diagram of Fenway Park, home of the Boston Red Sox. The park opened in 1912 and is still being used today. It too was expanded and remodeled over the years. The park's most distinctive feature is the left-field wall. It is called the Green Monster as a result of its paint job. Also, because it is so close to home plate, pitchers have grown to fear it.

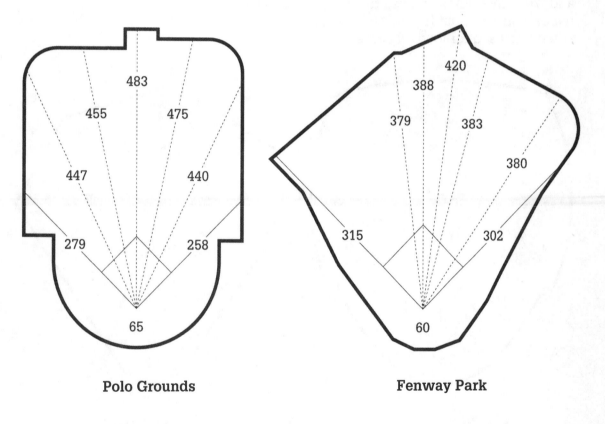

Polo Grounds   **Fenway Park**

Fields of Dreams (cont'd.)

This page shows Cleveland Stadium, once called Municipal Stadium, and Oriole Park at Camden Yards, located in Baltimore. The Cleveland park was built in 1932 in hopes that the Olympic Games would be awarded to Cleveland, but the games went elsewhere that year. The Indians did not play there on a regular basis until 1947. It is a huge, and essentially featureless, park, and as such it anticipated the bland, circular stadia constructed during the 1960s and '70s. The Indians moved to a new ballpark in 1994, but the old stadium is still being used for football games.

Oriole Park at Camden Yards opened in 1992 and was designed purposely to replicate the look and feel of such classic ballparks as the Polo Grounds and Fenway Park. An old warehouse was not only saved but renovated, and its roof was used to support a bank of floodlights for the stadium. The Orioles have enjoyed near-capacity crowds for every home game during the first two years of the park's existence, affirming, many believe, the value of a ballpark's "character."

From *Baseball Math: Grandslam Activities and Projects for Grades 4–8* published by Good Year Books. Copyright © 2001, 2005 Christopher Jennison.

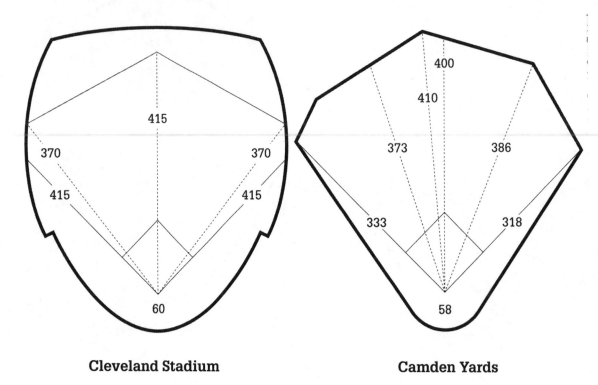

Cleveland Stadium　　　　　**Camden Yards**

You might undertake a research project that compared seasonal attendance figures for teams that moved from an old ballpark to a new one. What were the attendance figures for the Orioles during the years just prior to their move? Did the team play better in the new park? What effect might this have had on attendance figures?

Projects

Counting the Crowd

On the next page are the major-league attendance figures for the years 2002–2004.

Using the grid below, make a bar graph for three teams in each league.

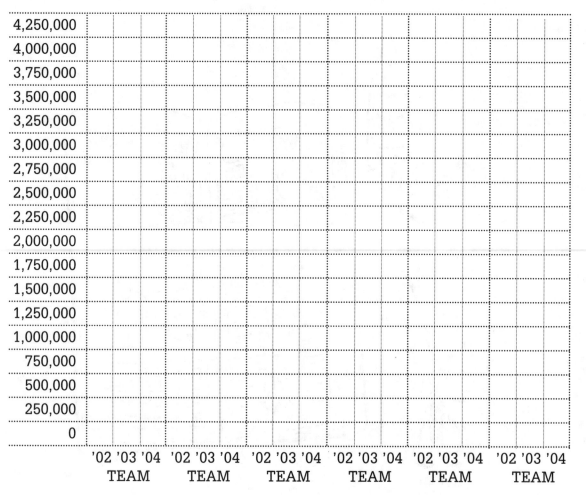

American League	2002	2003	2004	National League	2002	2003	2004
BAL	2,682,439	2,454,523	2,744,013	ATL	2,603,484	2,401,084	2,322,565
BOS	2,650,862	2,724,165	2,837,304	CHI	2,693,096	2,962,630	3,170,184
CAL	2,305,547	3,061,094	3,375,677	CIN	1,855,787	2,355,259	2,287,250
CHI	1,676,911	1,939,524	1,930,337	HOU	2,517,357	2,454,241	3,087,872
CLE	2,616,940	1,730,002	1,814,401	LA	3,131,255	3,138,626	3,488,283
DET	1,503,623	1,368,245	1,917,004	MON	812,045	1,025,639	748,550
KC	1,323,036	1,779,895	1,661,478	NY	2,804,838	2,140,599	2,318,321
MIL	1,969,153	1,700,354	2,062,382	PHI	1,618,467	2,259,948	3,206,532
MIN	1,924,473	1,946,011	1,879,222	PIT	1,784,988	1,636,751	1,583,031
NY	3,465,807	3,465,600	3,775,292	STL	3,011,756	2,910,386	3,048,427
OAK	2,169,811	2,216,596	2,201,516	SD	2,220,601	2,030,084	3,040,046
SEA	3,542,938	3,268,509	2,940,731	SF	3,253,203	3,138,626	3,258,864
TEX	2,352,397	2,094,394	2,513,685	TOT	28,306,877	28,453,873	31,556,925
TOR	1,637,900	1,799,458	1,900,041	ML	60,126,714	60,002.243	65,110,068
TOT	31,819,837	31,548,370	33,553,143				

1. Which league has the highest attendance figures?

_____

2. Which team showed the most consistent improvement over the three-year period?

_____

Can you think of one reason for this?

_____

Counting the Crowd (cont'd.)

3. Which team's attendance dropped most dramatically? Why do you think this happened?

_____

_____

_____

_____

_____

_____

4. Is the attendance trend up or down? Look at the graph and see if you can tell what the general tendency is from the heights of the bars and the directions they are indicating.

_____

_____

_____

_____

_____

5. What factors do you think have the most influence on attendance figures?

_____

_____

_____

_____

_____

_____

_____

_____

6. Do you think more people will attend games in future years, or do you think fewer people will? Why do you think so?

_____

_____

_____

_____

_____

_____

_____

_____

From Baseball Math: Grandslam Activities and Projects for Grades 4–8 published by Good Year Books. Copyright © 2001, 2005 Christopher Jennison.

Player Targets

An ultimate goal for major-league hitters is 3,000 hits during their entire careers. Very few players actually achieve this goal. Below is a list of several players with their hits totals through the 2004 season.

Craig Biggio 2639
Ivan Rodriguez......2051
Ken Griffey, Jr.........2156
Jeff Bagwell...........2289
Larry Walker..........2069
Jeff Kent 1910

1. Ken Griffey, Jr., had played 16 years at the end of the 2004 season. What is his average number of hits per season?

_____

2. Assuming he maintains this average, when will he reach the 3,000-hit total?

_____

3. Ken Griffey, Jr., was 35 years old when the 2005 season began, and his first major-league season was 1989. Assuming he maintains his average yearly hit total, how many more years will he have to play to reach 3,000 hits?

_____

4. Who was the most recent player to reach the 3,000-hit milestone? How many years did it take him to do it? Did he have a higher hit-per-year average in the first half of his career or in the second half? Why do you think this was so?

_____

_____

_____

_____

5. Who do you think will be the next player to reach 3,000 hits in his lifetime? Calculate his average number of hits per year.

_____

_____

6. Do you have a favorite player who is not listed in the group at the top of the page? Do you think he will reach 3,000 hits before he retires? Explain why.

_____

_____

_____

_____

7. A target for pitchers is 300 wins in a career. Twenty-two players have achieved this total so far. Cy Young, who pitched for 22 years from 1890–1911, won a total of 511 games! Between 1891 and 1904 he averaged slightly more than 24 wins per season. Today's pitchers don't stand a chance of winning 500 games, but a few could reach 300. Using the same calculations and estimates you made for hitters, list three pitchers who you feel will win 300 games by the end of their careers, and in each case write out your calculations.

_____

_____

_____

_____

_____

Time's Up

Some people think that baseball games are getting too long. There was a time, not so long ago, when games lasted between two and two and a half hours. Are games longer today? Below is a chart you can fill in over a twenty-game period. It will compare the lengths of games and if they were night or day games. The information you need is provided in your newspaper.

	Home	Away	Night	Day	Length			Home	Away	Night	Day	Length
1.	❑	❑	❑	❑	_____		11.	❑	❑	❑	❑	_____
2.	❑	❑	❑	❑	_____		12.	❑	❑	❑	❑	_____
3.	❑	❑	❑	❑	_____		13.	❑	❑	❑	❑	_____
4.	❑	❑	❑	❑	_____		14.	❑	❑	❑	❑	_____
5.	❑	❑	❑	❑	_____		15.	❑	❑	❑	❑	_____
6.	❑	❑	❑	❑	_____		16.	❑	❑	❑	❑	_____
7.	❑	❑	❑	❑	_____		17.	❑	❑	❑	❑	_____
8.	❑	❑	❑	❑	_____		18.	❑	❑	❑	❑	_____
9.	❑	❑	❑	❑	_____		19.	❑	❑	❑	❑	_____
10.	❑	❑	❑	❑	_____		20.	❑	❑	❑	❑	_____

1. What was the average length of the games?

   _____

2. Were your team's home games longer or shorter than games they played away?

   _____

3. Were night games longer or shorter?

   _____

4. Do you think baseball games are too long?

   _____

5. What could be done to shorten them?

   _____

   _____

To make this project more interesting, you could create another chart and compare American League and National League games to see if there is any difference in the lengths of games. On this chart you would only list the game lengths. Was there a significant difference (fifteen minutes or more) on average, from one league to the other? If there was a difference, can you think of one reason for the difference?

Not-So-Old-Timers

Lots of friendly arguments are caused when modern players are compared with so-called old timers. Below is a lineup of players who played during the 1950s and 1960s. Their best single-season records are given and their career records.

Best Season

Player	Pos.	AB	R	H	HR	RBI	AVG
Berra	C	584	88	179	22	125	.307
Musial	1B	611	135	230	39	131	.376
J. Robinson	2B	593	122	203	16	124	.342
B. Robinson	3B	612	82	194	28	118	.317
Banks	SS	617	119	193	47	129	.313
Williams	LF	566	150	194	43	159	.343
Mays	CF	580	123	185	51	127	.319
Aaron	RF	629	116	223	39	123	.355

Career

AB	R	H	HR	RBI	AVG
7555	1175	2150	358	1430	.285
10,972	1949	3630	475	1951	.331
4877	947	1518	137	734	.311
10,654	1232	2848	268	1357	.267
9421	1305	2583	512	1636	.274
7706	1798	2654	521	1839	.344
10,881	2062	3283	660	1903	.302
12,364	2174	3771	755	2297	.305

Best Season / Career

Player	Pos.	W	L	PCT.	G	IP	H	BB	SO	ERA	W	L	PCT.	G	IP	H	BB	SO	ERA
Koufax	P	27	9	.750	41	323	241	77	317	1.73	165	87	.655	397	2324	1754	817	2396	2.76

On the form on the next page, write the names of the players you think are the best at their positions today. List their best single-season records and then their estimated career totals. To make an estimate of a player's career totals, calculate their average statistics in each category per season, then multiply the average number by the number of seasons you think the player will remain in the majors. Remember that production usually drops in a player's last years, so you should probably use lower-than-average figures for a player's last two or three seasons.

1. How do your all-stars compare with the old timers?

_____

_____

_____

2. How many of your players were better?

_____

_____

_____

From Baseball Math: Grandslam Activities and Projects for Grades 4–8 published by Good Year Books. Copyright © 2001, 2005 Christopher Jennison.

		Best Season							Career					
Player	Pos.	AB	R	H	HR	RBI	AVG	AB	R	H	HR	RBI	AVG

		Best Season									Career								
Player	Pos.	W	L	PCT.	G	IP	H	BB	SO	ERA	W	L	PCT.	G	IP	H	BB	SO	ERA

3. If you were to create a team from both lineups, how many old timers would be on it?

_____

_____

Why?

_____

_____

4. Do you think modern players are better than the old timers?

_____

Why?

_____

_____

5. How many of baseball's most important records are held by players who played before 1970?

_____

_____

Challenge problem

Another measure of a pitcher's effectiveness that doesn't always appear in the record books is his number of strikeouts compared to the number of walks he gave up. During Sandy Koufax's best season, his strikeouts-to-walks ratio was better than four to one. What was his career strikeouts-to-walks ratio?

_____

Are there any modern players with ratios as good? What about Roger Clemens?

_____

Follow Your Favorite

Here's a chance to follow your favorite player over a long period. The chart on the next page provides enough space for sixteen weeks, but you can complete the project in less time, if you prefer. You'll keep a record of your favorite player's cumulative and weekly batting average, and then plot the information on a graph for easy reference. The sample chart below will help get you started. To figure out your player's weekly average, you first find the difference between last week's and this week's hits. Then you find the difference between last week's at-bats and this week's at-bats. Divide the number of hits by times at bat to get the weekly average.

| | | Cumulative | | | Weekly | |
Week	Hits	Times at bat	Avg.	Hits	Times at bat	Avg.
Week 1	7	21	.333	7	21	.333
Week 2	13	45	.289	13–7=6	45–21=24	.250
Week 3	20	66	.303	20–13=7	66–45=21	?

From *Baseball Math: Grandslam Activities and Projects for Grades 4–8* published by Good Year Books. Copyright © 2001, 2005 Christopher Jennison.

Stats for your favorite player: _____

| Week | Hits | Cumulative | | | Weekly | | |
		Times at bat	Avg.	Hits	Times at bat	Avg.

Follow Your Favorite (cont'd.)

Batting average for your favorite player: _____

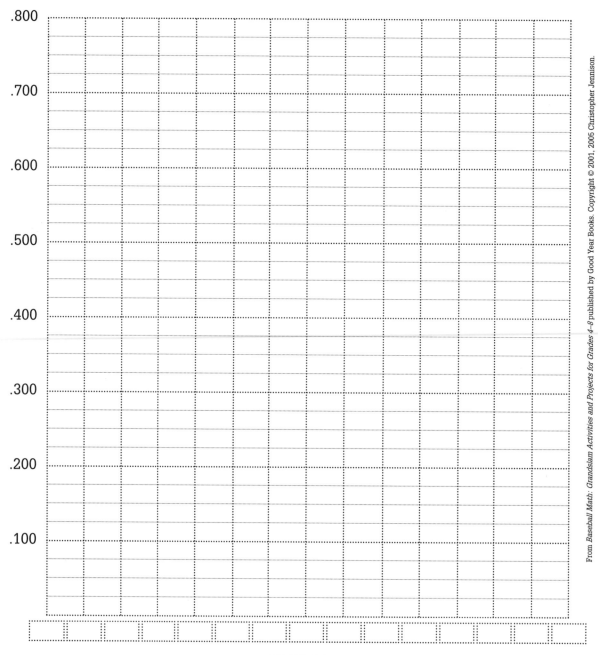

WEEK NO.

Keeping Score

Keeping score makes a baseball game a lot more fun, and it's not as complicated as it looks. First of all, each player position has a number—not a uniform number, but a number used for scorekeeping purposes. These numbers are always the same: 1-Pitcher, 2-Catcher, 3-First Baseman, 4-Second Baseman, 5-Third Baseman, 6-Shortstop, 7-Left Fielder, 8-Center Fielder, 9-Right Fielder. Below are marks and symbols used for keeping score.

Single .... ——	Reached on error..... E	Stolen base .... SB	Balk....................... BK
Double................. ══	Fielder's choice ..... FC	Sacrifice.............. SAC	Strikeout.................K
Triple .... ≡	Hit by Pitch...........HP	Sacrifice Fly .... SF	Walk .... BB
Homerun.............. ≣	Wild Pitch ....WP	Passed Ball ....PB	Force Out ....FO

Pretend that the scoring square or block is a baseball diamond, and that the lower-left corner is home plate, the lower-right corner is first base, the upper-right corner is second base, and the upper-left corner is third base. Below are four examples of scorekeeping:

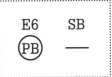

In this example, the hitter got to first on a single, stole second base, got to third on an error by the shortstop and scored on a passed ball. Put a circle around a scoring play so runs on a scorecard will show up at a glance.

In this example the batter hit a ground ball to the third baseman who threw to the first baseman for the putout.

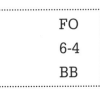

In this example, the batter reached first on a walk, and then was out at second on a force play when the shortstop fielded a ground ball hit by the next batter and threw to the second baseman who made the putout.

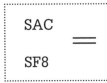

On this example, the batter hit a double, reached third on a sacrifice by the next batter, then scored when the next batter hit a sacrifice fly to the center fielder.

Keeping Score (cont'd.)

Take a look at three complete innings of scorekeeping. This will give you a better idea of how an entire game would look on a scorecard.

PLAYER	POS.	1	2	3	4	5	6	7	8	9	10	11
Wilson	2B	K		E̲								
O'Brien	LF	9		6-4 E5								
Chavez	3B	⬓	6-4-3									
Cole	1B	3-1		///								
Nieman	C	///	2									
Bevington	CF		6-3									
Dempsey	RF		2-6̲									
Lubin	SS		///	3								
Gamboli	P			⊖̿								
TOTALS HITS / RUNS		1 / 1	0 / 1	/								

From *Baseball Math: Grandslam Activities and Projects for Grades 4–8* published by Good Year Books. Copyright © 2001, 2005 Christopher Jennison.

In the first inning, Wilson struck out, O'Brien hit a fly ball to the right fielder, and then Chavez hit a home run. Cole ended the inning with a groundball to the first baseman, who threw to the pitcher covering first for the out. Be sure to make a mark in the box below so you don't start there in the second inning. Nieman started the second inning by popping out to the catcher, then Bevington hit a groundball to the shortstop who threw to first for the out. Dempsey followed with a single, but was out stealing when the catcher threw to the shortstop covering second base. Lubin began the third inning by popping out to the first baseman, then Gamboli doubled and scored on Wilson's single. O'Brien reached first base on an error by the third baseman and Wilson got to second on the same error. Chavez then hit into a double play. The shortstop fielded

his grounder, threw to the second baseman for one out, then the second baseman threw to first for the second out. O'Brien was out at second.

Now you're ready to score a game. Use the scoresheet below. You'll need a separate sheet for each team. Score a game that you listen to on the radio. Announcers know that many listeners keep score and they describe the action with that in mind.

Challenge problem

You can total each player's performance in the columns on the right of the card. Find out how many official at-bats and hits one player had before the game, then combine what the player did in this game with those figures. You'll be able to compute his new batting average before it appears in tomorrow's newspaper! Make similar calculations for the teams' places in the standings.

TEAM:

PLAYER	POS.	1	2	3	4	5	6	7	8	9	10	11	RUNS	HITS	RBI	A	PO	E
	2B																	
	LF																	
	3B																	
	1B																	
	C																	
	CF																	
	RF																	
	SS																	
	P																	
TOTALS	H / R																	

H=hits R=runs

From *Baseball Math: Grandslam Activities and Projects for Grades 4–8* published by Good Year Books. Copyright © 2001, 2005 Christopher Jennison.

Above Average

Every Sunday during the baseball season, the sports sections of most newspapers print player and team statistics. *USA Today* prints American League statistics every Tuesday and National League statistics every Wednesday. Statistics for both leagues appear in *Baseball Weekly*. It's easy to see who the top players are and which teams are leading their divisions. But do the "star" players always play for the league-leading teams? Take a look at team batting averages, for instance. Are the best hitting teams leading their divisions? What about team earned run averages, the statistic used to measure a team's pitching strength? What about fielding averages? Perhaps some

combination of these statistics would provide a better measure of a team's effectiveness. One simple computation would be to add a team's batting and fielding averages, divide by two, and subtract the team earned run average. Do this calculation for several teams. Does this "new" statistic provide a more accurate measure of a team's ability?

Using the calculation suggested above and the chart below, follow the progress of these teams over a period of time. You could do this for any time period you prefer, but the longer you follow the teams, the more reliable the information will be.

TEAMS	WEEK 1		WEEK 2		WEEK 3		WEEK 4		WEEK 5		WEEK 6	
	Combined Average	Division Standing	Combined Average	Division Standing	Combined Average	Division Standing	Combined Average	Division Standing	Combined Average	Division Standing	Combined Average	Division Standing

Was there any relationship between a team's combined statistics and their position in the division standings?

_____

_____

Can you name other factors that determine a team's success?

_____

_____

From *Baseball Math: Grandslam Activities and Projects for Grades 4–8* published by Good Year Books. Copyright © 2001, 2005 Christopher Jennison.

On page 70 is a diagram of a baseball stadium, surrounded by streets and city blocks. Many ballparks are located in inner-city areas, and their designs were carefully created to make maximum use of minimum space.

Now you have an opportunity to design your own ballpark, and add, if you wish, an imaginary surrounding neighborhood. Begin your design by drawing the baseball diamond and the foul lines. The foul lines extend from home plate at a 90° angle. The minimum distance from home plate to the foul pole in a major league stadium is 325 feet, but you can make it longer if you want to. (The batters on your team will not appreciate too long a distance, however.) The minimum distance from home plate to straight-away center field is 400 feet, but this can be extended as well. Also, the minimum distance from home plate to the grandstand behind home plate is 60 feet. The distance between home plate and first base, and between all the other bases is 90 feet. Finally, the distance between home plate and the middle of the pitcher's mound is 60 feet, 6 inches.

Obviously your drawing can't show the actual distances, so you will have to use a scale. As a suggestion, use 1 inch for every 90 feet. That will simplify the measurements between bases. How many inches would you use if you wanted the distance from home plate to each foul pole to be 360 feet?

How many inches would you use if you wanted the distance from home plate to straight-away center field to be 405 feet?

After you have drawn the diamond and the foul lines, you can draw the grandstand and the outfield fences. The outfield fences can be curved, or shaped in almost any angle you can imagine, within reason. Make sure the minimum distances from home plate to the foul poles and center field are followed as described above. The grandstand can also be any shape you want it to be. The drawing on page 70 is only a suggestion. You can add as many details as you like, including dugouts, bullpens, seats, and lights.

To make your design even more interesting, you can draw streets and city blocks such as those shown on the drawing. You might want to include a parking lot near your ballpark. Details will make your drawing more interesting. Think about including subway stops, names of buildings, and names of streets. And think about a name for your ballpark. You might want to name it after yourself; after all, you did all the work designing it!

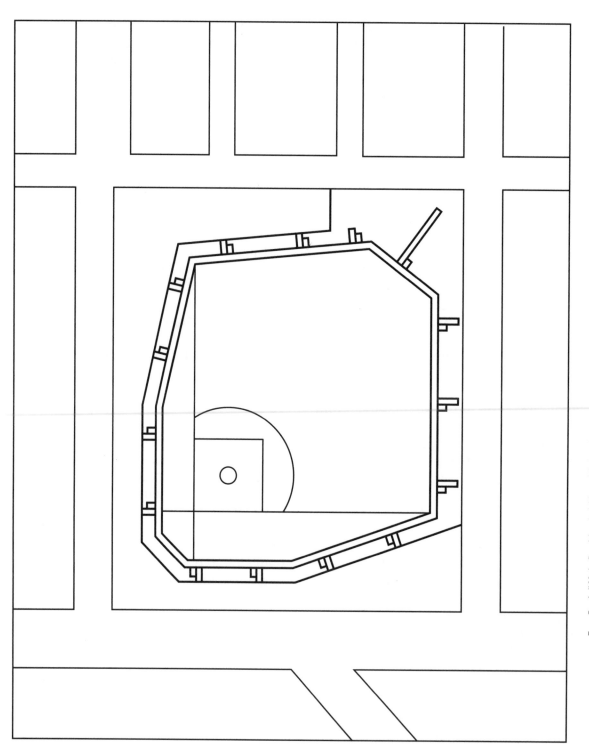

From *Baseball Math: Grandslam Activities and Projects for Grades 4–8* published by Good Year Books. Copyright © 2001, 2005 Christopher Jennison.

Lopsided Ballpark

When the Brooklyn Dodgers left Brooklyn and became the Los Angeles Dodgers in 1958, they had to play for four seasons in the Los Angeles Coliseum before their new ballpark was built. The Coliseum was a perfect shape for football games and track meets, which it was designed for, but it was a poor place to play baseball. Take a look at the diagram below. It shows the distance from home plate to the foul poles and the outfield fence. As you can see, the distances to the foul poles are very short; in fact, the distance to the left field foul pole is the minimum distance permitted by the rules. A wire mesh screen, 42 feet high, was put up in front of the left field grandstand from the foul pole to a point 320 feet from home plate. This was done to keep pop flys from becoming homers, but in the first nine games played in the Coliseum 20 homers were hit over the screen. And although the right field foul pole is close to home plate, the right field fence actually started about 370 feet from home, which made it very hard for left-handed hitters to hit homers.

Maybe you can do a better job designing the playing field. The dimensions of the Coliseum will stay the same, of course, but could you reposition the playing area so that the distances down the foul lines are more identical? Begin by making a tracing of the Coliseum dimensions. Use the same scale (1 in. = 90 ft.) for drawing the diamond, and then see how far out the foul lines extend until they meet the grandstand. Are they at least the minimum distance of 250 feet in each case?

1. Do you think this is a better arrangement? Why?

2. Would you put up screens in front of the grandstands? Why or why not?

3. What distances would you set for the outfield fence? Why?

4. Why do you think the owners of the Dodgers arranged the playing field the way they did?

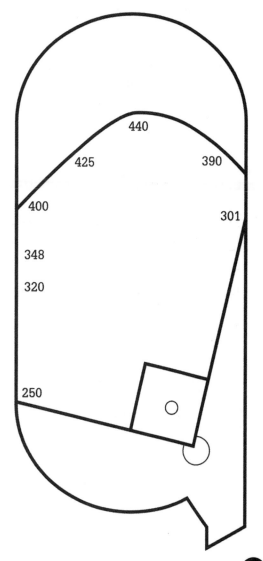

Star Wars

It is fun to compare old-time players with players from recent years and try to figure out which ones were better. This project will not decide the issue, but it will give you a chance to make some comparisons and may enable you to strengthen your argument.

On the following eight pages are the names and career records of all players admitted to baseball's Hall of Fame through 2005. With two or three friends, select two all-star teams. One team should be made up of players born before 1925. The other team should include players born from 1925 and beyond. Select twenty-five players for each team. Ten players on each team should be pitchers.

Now select starting teams from each roster. For your starting teams compare the following:

1. A composite batting average for all players except the pitchers. (This is the average of the player's batting averages.)
2. The yearly home run averages for all players except pitchers.
3. The yearly RBI averages for all players except pitchers.
4. The yearly runs scored averages for all players except pitchers.
5. The yearly average of games won by each pitcher.
6. A composite earned run average for each pitcher.
7. A composite won-lost percentage for each pitcher.
8. The yearly average of shutouts by each pitcher.

Because baseball always involves more players than those on the starting team, make the same comparisons for the rest of the players on your squads. Based on these comparisons, do you want to make any changes in your starting lineups?

_____

Which starting lineup had the better statistics?

_____

Which overall team had the better statistics?

_____

To make the comparison, you can come up with team averages in batting and pitching.

Now you have some basis for comparing old-timers with more modern players. Are the modern players better?

Can you think of any other points to compare in deciding which players are better? Was baseball different during the early years of the century? In what way was it different, and how did it affect the players' performances?

To study these players and the years in which they played, you can read any one of a number of baseball history books in your school or local library. Some good ones include:

Alexander, Charles C. *Our Game: An American Baseball History.* New York: Holt, 1997.

Dewey, Donald and Acocella, Nicholas. *The Biographical History of Baseball, Revised Edition.* Chicago: Triumph Books, 2003.

Neyer, Rob. *Rob Neyer's Big Book of Baseball Lineups.* New York: Fireside, 2003.

Okrent, Daniel, ed. *The Ultimate Baseball Book.* Rev. ed. Boston: Houghton Mifflin, 2000.

Pietrusza, David, et al. *Baseball: The Biographical Encyclopedia.* New York: Total Sports, 2000.

Seaver, Tom. *The Old Ballgame: The Greatest Memories of Baseball's Greats.* New York: Doubleday, 2005.

Thorn, John et al. *Total Baseball: The Ultimate Baseball Encyclopedia, 8th edition.* Toronto: Sport Classic Books, 2004.

FIRST BASEMEN	YRS	B-T	HT.	WT.	BORN	G	AB	R	H	2B	3B	HR	RBI	PCT
Anson, Cap	22	R-R	6-1	227	4/17/1852	2253	9084	1712	3081	530	129	92	1880	.339
Beckley, Jake	20	L-L	6-1	180	8/4/1867	2373	9476	1601	2930	455	46	87	1575	.309
Bottomley, Jim	16	L-L	6-0	180	4/23/1900	1991	7471	1177	2313	465	151	219	1422	.310
Brouthers, Dan	19	L-L	6-2	207	5/8/1858	1658	6725	1507	2349	446	212	103	1296	.349
Cepeda, Orlando	17	R-R	6-2	210	9/19/1974	2124	7927	1131	2351	417	27	379	1365	.297
Chance, Frank	17	R-R	6-0	190	9/9/1877	1232	4279	796	1273	195	80	20	405	.297
Connor, Roger	18	L-L	6-2	210	7/1/1857	1987	7807	1607	2535	429	227	132	1296	.325
Foxx, Jimmie	20	R-R	6-0	190	10/22/1907	2317	8134	1751	2646	458	125	534	1921	.325
Gehrig, Lou	17	L-L	6-1	212	6/19/1903	2164	8001	1888	2721	534	163	493	1990	.340
Greenberg, Hank	13	R-R	6-4	218	1/1/1911	1394	5193	1051	1628	379	71	331	1276	.313
Kelly, George	16	R-R	6-3	190	9/10/1896	1622	5993	8191	1778	337	76	148	1019	.297
Killebrew, Harmon	22	R-R	6-0	210	6/29/1936	2435	8147	1283	2086	290	24	573	1584	.256
McCovey, Willie	22	L-L	6-4	225	1/10/1938	2588	8197	1229	2211	353	46	521	1555	.270
Mize, Johnny	15	L-R	6-2	215	1/7/1913	1884	6443	1118	2011	367	83	359	1337	.312
Murray, Eddie	21	Bo-R	6-2	200	2/24/1956	3026	11,336	1627	3255	560	35	504	1917	.287
Sisler, George	15	L-L	5-10½	170	3/24/1893	2055	8267	1284	2812	425	164	102	1180	.340
Stargell, Willie	21	L-L	6-3	220	3/6/1941	2360	7927	1195	2232	423	55	475	1540	.282
Terry, Bill	14	L-L	6-0½	200	10/30/1898	1721	6428	1120	2193	373	112	154	1078	.341

Star Wars (cont'd.)

SECOND BASEMEN	YRS	B-T	HT.	WT.	BORN	G	AB	R	H	2B	3B	HR	RBI	PCT
Carew, Rod	19	L-R	6-0	182	10/1/1945	2469	9315	1424	3053	445	112	92	1015	.328
Collins, Eddie	25	L-R	5-9	175	5/2/1887	2826	9946	1816	3309	437	186	47	1307	.333
Doerr, Bobby	14	R-R	5-11	185	4/7/1918	1865	7093	1094	2042	381	89	223	1247	.288
Evers, Johnny	18	L-R	5-9	140	7/21/1881	1776	6136	919	1659	216	70	12	538	.270
Fox, Nelson	19	L-R	5-9	150	12/25/1927	2367	9232	1279	2663	355	112	35	790	.288
Frisch, Frankie	19	Bo-R	5-10	185	9/9/1898	2311	9112	1532	2880	466	138	105	1242	.316
Gehringer, Charlie	19	L-R	5-11½	185	5/11/1903	2323	8860	1774	2839	574	146	184	1427	.320
Herman, Billy	15	R-R	5-11	195	7/7/1909	1922	7707	1163	2345	486	82	47	839	.304
Hornsby, Rogers	23	R-R	5-11	200	4/27/1869	2259	8173	1579	2930	541	169	301	1579	.358
Lajoie, Nap	21	R-R	6-1	195	10/5/1875	2475	9589	1506	3252	652	164	82	1599	.339
Mazeroski, Bill	17	R-R	5-10	183	9/5/1936	2163	7755	769	2016	294	62	138	853	.260
Molitor, Paul	21	R-R	6-0	185	8/22/1956	2683	10,835	1782	3319	605	114	234	1307	.306
Morgan, Joe	22	L-R	5-10	170	6/29/1926	2649	9277	1650	2517	449	96	268	1133	.271
Robinson, Jackie	10	R-R	6-0	225	1/31/1919	1382	4877	947	1518	273	54	137	734	.311
Sandberg, Ryne	16	R-R	6-2	180	9/18/1959	2164	8385	1318	2386	403	76	282	1061	.285
Schoendienst, Al	19	Bo-R	6-0	170	2/2/1923	2216	8479	1223	2449	427	78	84	773	.289

SHORTSTOPS	YRS	B-T	HT.	WT.	BORN	G	AB	R	H	2B	3B	HR	RBI	PCT
Aparicio, Luis	18	R-R	5-8	156	4/29/1934	2599	10,230	1335	2677	394	92	83	791	.262
Appling, Luke	20	R-R	5-11	200	4/2/1907	2422	8856	1319	2749	440	102	45	1116	.310
Bancroft, Dave	16	Bo-R	5-9	160	4/20/1892	1913	7182	1048	2004	320	77	32	579	.279
Banks, Ernie	19	R-R	6-1	180	1/31/1931	2528	9421	1305	2583	407	90	512	1636	.274
Boudreau, Lou	15	R-R	5-11	193	7/17/1917	1646	6029	861	1779	385	66	68	789	.295
Cronin, Joe	20	R-R	5-11½	180	10/12/1906	2124	7579	1233	2285	516	118	170	1423	.301
Jackson, Travis	15	R-R	5-10½	160	11/2/1903	1656	6086	833	1768	291	86	135	929	.291
Jennings, Hugh	17	R-R	5-8½	165	4/2/1870	1264	4840	989	1520	227	88	19	840	.314
Maranville, Rabbit	23	R-R	5-5	155	11/11/1891	2670	10,078	1255	2605	380	177	28	874	.258
Reese, Pee Wee	16	R-R	5-9½	178	7/23/1918	2166	8058	1338	2170	330	80	126	885	.269
Rizzuto, Phil	13	R-R	5-6	160	9/25/1917	1661	5816	877	1588	239	62	38	563	.273
Sewell, Joe	14	L-R	5-7	160	10/9/1898	1903	7132	1141	2226	436	68	49	1053	.312
Smith, Ozzie	19	Bo-R	5-11	150	12/26/1954	2573	9396	1257	2460	402	69	28	793	.262
Tinker, Joe	15	R-R	5-9	175	7/27/1880	1642	5936	716	1565	238	106	29	783	.264
Vaughan, Arky	14	L-R	5-11	185	3/9/1912	1817	6622	1173	2103	356	128	96	926	.318
Wagner, Honus	21	R-R	5-11	200	2/24/1874	2785	10,427	1740	3430	651	252	101	1732	.329
Wallace, Bobby	25	R-R	5-8	170	11/4/1874	2369	8629	1056	2308	394	149	36	1121	.267
Ward, Monte	17	L-R	5-9	165	3/3/1860	1810	7579	1403	2151	232	95	26	605	.283
Yount, Robin	20	R-R	6-0	170	9/16/1955	2856	11,008	1632	3142	583	126	251	1406	.285

From *Baseball Math: Grandslam Activities and Projects for Grades 4–8* published by Good Year Books. Copyright © 2001, 2005 Christopher Jennison.

THIRD BASEMEN	YRS	B-T	HT.	WT.	BORN	G	AB	R	H	2B	3B	HR	RBI	PCT
Baker, Frank	13	L-R	5-11½	180	3/13/1886	1575	5983	887	1838	313	103	96	1012	.307
Boggs, Wade	18	L-R	6-2	197	6/15/1958	2440	9180	1513	3010	578	61	118	1014	.328
Brett, George	21	L-R	6-0	200	5/15/1953	2707	10,349	1583	3154	665	137	317	1595	.305
Collins, Jimmy	14	R-R	5-8	160	1/16/1873	1718	6792	1057	1999	333	117	62	985	.294
Kell, George	15	R-R	5-10	170	8/23/1922	1795	6702	881	2054	385	50	78	870	.306
Killebrew, Harmon	22	R-R	5-11	213	6/29/1936	2435	8147	1283	2086	290	24	573	1584	.256
Lindstrom, Fred	13	R-R	5-11½	170	11/21/1905	1438	5611	895	1747	301	81	103	779	.311
Mathews, Eddie	17	L-R	6-1	200	10/13/1931	2391	8537	1509	2315	354	72	512	1453	.271
Robinson, Brooks	23	R-R	6-1	190	5/18/1937	2896	10,654	1232	2848	482	68	268	1357	.267
Schmidt, Mike	18	R-R	6-2	203	9/27/1949	2404	8352	1506	2234	408	59	548	1595	.267
Traynor, Pie	17	R-R	6-1	175	11/11/1899	1941	7559	1183	2416	371	164	58	1273	.320

LEFT FIELDERS	YRS	B-T	HT.	WT.	BORN	G	AB	R	H	2B	3B	HR	RBI	PCT
Brock, Lou	19	L-L	5-11½	172	6/18/1939	2616	10,332	1610	3023	486	141	149	900	.293
Burkett, Jesse	16	L-L	5-8	155	2/12/1870	2063	8389	1708	2872	314	185	72	952	.342
Clarke, Fred	21	L-R	5-10½	165	10/3/1872	2204	8584	1620	2703	358	219	65	1015	.315
Delahanty, Ed	16	R-R	5-10	170	10/31/1867	1825	7493	1596	2593	508	182	98	1464	.346
Goslin, Goose	18	L-R	5-11	180	10/16/1900	2287	8656	1483	2735	500	173	248	1609	.316
Hafey, Chick	13	R-R	6-1	185	2/12/1904	1283	4625	777	1466	341	67	164	833	.317
Irvin, Monte	8	R-R	6-1	195	2/25/1919	764	2499	366	731	97	31	99	443	.293
Kelley, Joe	17	R-R	5-11	190	12/9/1871	1829	6989	1425	2245	253	189	66	1194	.321
Kiner, Ralph	10	R-R	6-2	195	10/27/1922	1472	5205	971	1451	216	39	369	1015	.279
Manush, Heinie	17	L-L	6-0	200	7/20/1901	2008	7654	1287	2524	491	160	110	1183	.330
Medwick, Joe	17	R-R	5-10	185	11/4/1911	1984	7635	1198	2471	540	113	205	1383	.324
Musial, Stan	22	L-L	6-0	180	11/21/1920	3026	10,972	1949	3630	725	177	475	1951	.331
O'Rourke, Jim	19	R-R	5-8	185	8/24/1852	1750	7365	1425	2314	385	139	49	1010	.314
Simmons, Al	20	R-R	6-0	210	5/22/1902	2215	8761	1507	2927	539	149	307	1827	.334
Wheat, Zack	19	L-L	5-10	170	5/23/1888	2406	9106	1289	2884	476	172	132	1265	.317
Williams, Billy	18	L-R	6-1	175	6/15/1938	2488	9350	1410	2711	434	88	426	1475	.290
Williams, Ted	19	L-R	6-4	198	8/30/1918	2292	7706	1798	2654	525	71	521	1839	.344
Winfield, Dave	22	R-R	6-6	220	10/3/1951	2973	11,003	1669	3110	540	88	465	1833	.283
Yastrzemski, Carl	23	L-R	5-11	185	8/22/1939	3308	11,988	1816	3419	646	59	452	1844	.285

Star Wars (cont'd.)

CENTER FIELDERS	YRS	B-T	HT.	WT.	BORN	G	AB	R	H	2B	3B	HR	RBI	PCT
Ashburn, Richie	15	L-R	5-10	170	3/19/1927	2189	8365	1322	2574	317	109	29	586	.308
Averill, Earl	13	L-R	5-9½	170	5/21/1902	1669	6352	1224	2019	401	128	238	1164	.318
Carey, Max	20	Bo-R	6-0	166	1/11/1890	2469	9363	1545	2665	419	159	69	797	.285
Cobb, Ty	24	L-R	6-1	175	12/18/1886	3033	11,429	2245	4191	724	298	117	1960	.367
Combs, Earle	12	L-R	6-0	185	5/14/1899	1455	5746	1186	1866	309	154	58	629	.325
DiMaggio, Joe	13	R-R	6-2	195	11/25/1914	1736	6821	1390	2214	389	131	361	1537	.325
Duffy, Hugh	17	R-R	5-7	168	11/26/1866	1722	6999	1545	2307	310	117	103	597	.330
Hamilton, Billy	14	L-R	5-6	165	2/16/1866	1578	6262	1690	2157	225	94	37	736	.344
Mantle, Mickey	18	Bo-R	6-0	200	10/20/1931	2401	8102	1677	2415	344	72	536	1509	.298
Mays, Willie	22	R-R	5-11	187	5/6/1931	2992	10,881	2062	3283	523	140	660	1903	.302
Puckett, Kirby	12	R-R	5-8	210	3/14/1961	1783	7244	1071	2304	414	57	207	1085	.318
Roush, Edd	18	L-L	5-11	175	5/8/1893	1748	6646	1000	2158	311	168	63	882	.325
Snider, Duke	18	L-R	6-0	200	9/19/1926	2143	7161	1259	2116	358	85	407	1333	.295
Speaker, Tris	22	L-L	5-11½	193	4/4/1888	2791	10,200	1881	3516	793	222	117	1562	.345
Waner, Lloyd	18	L-R	5-9	145	3/16/1906	1993	7772	1201	2459	281	118	27	598	.316
Wilson, Hack	12	R-R	5-6	195	4/26/1900	1348	4760	884	1461	266	67	244	1062	.307

RIGHT FIELDERS	YRS	B-T	HT.	WT.	BORN	G	AB	R	H	2B	3B	HR	RBI	PCT
Aaron, Hank	23	R-R	6-0	190	2/5/1934	3298	12,364	2174	3771	624	98	755	2297	.305
Clemente, Roberto	18	R-R	5-11	182	8/18/1934	2433	9454	1416	3000	440	166	240	1305	.317
Crawford, Sam	19	L-L	6-0	190	4/18/1880	2505	9579	1392	2964	455	312	95	1525	.309
Cuyler, Kiki	18	R-R	5-10½	180	8/30/1899	1879	7161	1305	2299	394	157	128	1065	.321
Flick, Elmer	13	L-R	5-9	168	1/11/1876	1480	5601	951	1767	268	170	46	756	.315
Heilmann, Harry	17	R-R	6-1	200	8/3/1894	2145	7787	1291	2660	542	151	183	1552	.342
Hooper, Harry	17	L-R	5-10	165	8/24/1887	2308	8784	1429	2466	389	160	75	813	.281
Jackson, Reggie	21	L-L	6-0	200	4/18/1946	2820	9864	1551	2584	463	49	563	1702	.262
Kaline, Al	22	R-R	6-2	184	12/19/1934	2834	10,116	1622	3007	498	75	399	1583	.297
Keeler, Willie	19	L-L	5-4½	140	3/13/1872	2124	8564	1720	2955	234	155	32	810	.345
Kelly, King	16	R-R	5-11	180	12/31/1857	1434	5922	1359	1853	351	109	65	950	.313
Klein, Chuck	17	L-R	6-0	195	10/7/1905	1753	6486	1168	2076	398	74	300	1201	.320
McCarthy, Tommy	13	R-R	5-7	170	7/24/1864	1258	5055	1050	1485	195	57	43	732	.294
Ott, Mel	22	L-R	5-9	165	3/2/1909	2730	9456	1859	2876	488	72	511	1860	.304
Rice, Sam	20	L-L	5-10	150	2/20/1890	2404	9269	1514	2987	498	184	34	1077	.322
Robinson, Frank	21	R-R	6-1	194	8/31/1935	2808	10,006	1829	2943	528	72	586	1812	.294
Ruth, Babe	22	L-L	6-2	215	2/6/1895	2503	8397	2174	2873	506	136	714	2204	.342
Slaughter, Enos	19	L-R	5-9	190	4/27/1916	2380	7946	1247	2383	413	148	169	1304	.300
Thompson, Sam	15	L-L	6-2	207	3/5/1860	1405	6004	1259	2016	326	146	126	1299	.336
Waner, Paul	20	L-L	5-8	148	4/16/1903	2549	9459	1627	3152	605	191	113	1309	.333
Youngs, Ross	10	Bo-R	5-8	162	4/10/1897	1211	4627	812	1491	236	93	42	596	.322

Star Wars (cont'd.)

CATCHERS	YRS	B-T	HT.	WT.	BORN	G	AB	R	H	2B	3B	HR	RBI	PCT
Bench, Johnny	17	R-R	6-1	210	12/7/1947	2158	7658	1091	2048	381	24	389	1376	.267
Berra, Yogi	19	L-R	5-7½	190	5/12/1925	2120	7555	1175	2150	321	49	358	1430	.285
Bresnahan, Roger	17	R-R	5-8	180	6/11/1879	1410	4480	684	1251	222	72	26	531	.279
Campanella, Roy	10	R-R	5-9½	205	11/19/1921	1215	4205	627	1161	178	18	242	856	.276
Carter, Gary	19	R-R	6-2	215	4/8/1954	2296	7971	1025	2092	371	31	324	1225	.262
Cochrane, Mickey	13	L-R	5-10½	180	4/6/1903	1482	5169	1041	1652	333	64	119	832	.320
Dickey, Bill	17	L-R	6-1½	185	6/6/1907	1789	6300	930	1969	343	72	202	1209	.313
Ewing, Buck	18	R-R	5-10	188	10/27/1859	1281	5348	1118	1663	237	179	66	883	.313
Ferrell, Rick	18	R-R	5-11	170	10/12/1905	1884	6028	687	1692	324	45	28	734	.281
Fisk, Carlton	24	R-R	6-2	220	12/26/1947	2499	8756	1276	2356	421	47	376	1330	.269
Hartnett, Gabby	20	R-R	6-1	218	12/20/1900	1990	6432	867	1912	396	64	236	1179	.297
Lombardi, Ernie	17	R-R	6-3	230	4/6/1908	1853	5855	601	1792	277	27	190	990	.306
Schalk, Ray	18	R-R	5-7	154	8/12/1892	1760	5306	579	1345	199	48	12	596	.253

From *Baseball Math: Grandslam Activities and Projects for Grades 4–8* published by Good Year Books. Copyright © 2001, 2005 Christopher Jennison.

PITCHERS	YRS	B-T	HT.	WT.	BORN	G	IP	SHO	W	L	PCT	H	SO	BB	ERA
Alexander, Grover	20	R-R	6-1	185	2/26/1887	696	5188	90	373	208	.642	4868	2198	951	2.56
Bender, Chief	16	R-R	6-2	185	5/5/1884	433	2847	41	208	112	.650	2455	1630	667	2.46
Brown, Mordecai	14	Bo-R	5-10	175	10/19/1876	411	2697	55	208	111	.652	2284	1166	548	2.06
Bunning, Jim	17	R-R	6-3	195	10/23/1931	591	3760	40	224	184	.549	3433	2855	1000	3.27
Carlton, Steve	24	L-L	6-4	210	12/22/1944	741	5217	55	329	244	.574	4672	4136	1833	3.22
Chesbro, Jack	11	R-R	5-9	180	6/5/1874	392	2886	35	198	128	.607	2602	1276	674	2.68
Clarkson, John	12	R-R	5-10	160	7/1/1861	519	4514	37	327	176	.650	4384	2013	1192	2.81
Coveleski, Stan	14	R-R	5-9½	175	7/13/1890	450	3081	38	215	141	.604	3055	981	802	2.88
Dean, Dizzy	12	R-R	6-3	202	1/16/1911	317	1966	27	150	83	.644	1921	1155	458	3.04
Drysdale, Don	14	R-R	6-6	208	7/23/1936	518	3432	49	209	166	.557	3084	2486	855	2.95
Eckersley, Dennis	24	R-R	6-2	190	10/3/1954	1071	3286	20	197	171	.535	3076	2401	738	3.50
Faber, Red	20	Bo-R	6-1	190	9/6/1888	669	4089	29	254	212	.545	4104	1471	1213	3.15
Feller, Bob	18	R-R	6-0	185	11/3/1918	570	3828	46	266	162	.621	3271	2581	1764	3.25
Fingers, Rollie	17	R-R	6-4	195	8/25/1946	944	1701	2	114	118	.491	1474	1299	492	2.90
Ford, Whitey	16	L-L	5-10	180	10/21/1928	498	3171	45	236	106	.690	2766	1956	1086	2.74
Galvin, Pud	14	R-R	5-8	190	12/25/1856	697	5959	57	361	309	.539	6334	1786	744	2.87
Gibson, Bob	17	R-R	6-1	193	11/9/1935	528	3885	56	251	174	.591	3279	3117	1336	2.91
Gomez, Lefty	14	L-L	6-2	175	11/26/1908	368	2503	28	189	102	.649	2290	1468	1095	3.34
Grimes, Burleigh	19	R-R	5-10	195	8/18/1893	615	4178	35	270	212	.560	4406	1512	1295	3.52
Grove, Lefty	17	L-L	6-3	204	3/6/1900	616	3940	35	300	141	.680	3849	2266	1187	3.06
Haines, Jesse	19	R-R	6-0	180	7/22/1893	555	3208	23	210	158	.571	3460	981	871	3.64
Hoyt, Waite	21	R-R	5-11	183	9/9/1899	675	3762	26	237	182	.566	4037	1206	1003	3.59
Hubbell, Carl	16	R-L	6-1	172	6/22/1903	535	3591	36	253	154	.622	3461	1677	725	2.98
Hunter, Catfish	15	R-R	6-0	190	4/18/1946	500	3449	42	224	166	.574	2958	2012	954	3.26
Jenkins, Ferguson	19	R-R	6-5	210	12/14/1943	664	4500	49	284	226	.557	4142	3192	977	3.34
Johnson, Walter	21	R-R	6-1	200	11/6/1887	802	5923	113	416	279	.599	4920	3508	1353	2.17
Joss, Addie	9	R-R	6-3	185	4/12/1880	286	2329	45	160	97	.623	1893	915	357	1.89
Keefe, Tim	14	R-R	5-10½	185	1/1/1857	598	5043	40	342	224	.604	4524	2538	1225	2.63
Koufax, Sandy	12	R-L	6-2	202	12/30/1935	397	2325	40	165	87	.655	1754	2396	817	2.76
Lemon, Bob	13	L-R	6-0	180	9/22/1920	460	2849	31	207	128	.618	2559	1277	1251	3.23
Lyons, Ted	21	Bo-R	5-11	200	12/28/1900	594	4162	27	260	230	.531	4489	1073	1121	3.67
Marichal, Juan	16	R-R	5-11	190	10/20/1938	471	3506	52	243	142	.631	3153	2303	709	2.89
Marquard, Rube	18	Bo-L	6-3½	175	10/9/1889	536	3307	30	201	177	.532	3233	1593	858	3.13

From *Baseball Math: Grandslam Activities and Projects for Grades 4–8* published by Good Year Books. Copyright © 2001, 2005 Christopher Jennison.

Star Wars (cont'd.)

PITCHERS	YRS	B-T	HT.	WT.	BORN	G	IP	SHO	W	L	PCT	H	SO	BB	ERA
Mathewson, Christy	17	R-R	6-1½	195	8/12/1870	635	4781	83	373	188	.665	4203	2505	837	2.13
McGinnity, Joe	10	R-R	5-11	206	3/19/1871	467	3455	32	247	145	.630	3235	1065	799	2.66
Nichols, Kid	15	R-R	5-10½	175	9/14/1869	620	5067	48	361	208	.634	4854	1866	1245	2.96
Niekro, Phil	24	R-R	6-1	180	4/1/1939	864	5404	45	318	274	.537	5044	3342	1809	3.35
Paige, Satchel	6	R-R	6-4	180	7/7/1906	179	476	4	28	31	.475	429	290	183	4.09
Palmer, Jim	19	R-R	6-3	196	10/15/1945	558	3948	53	268	152	.638	3349	2212	1311	2.86
Pennock, Herb	22	R-R	6-0	165	2/10/1894	617	3572	35	240	162	.597	3900	1227	916	3.60
Perry, Gaylord	22	Bo-L	6-4	215	9/15/1938	777	5350	53	314	265	.542	4938	3534	1379	3.11
Plank, Eddie	17	L-L	5-11½	175	8/31/1875	581	4234	69	305	181	.628	3688	2112	984	2.35
Radbourn, Old Hoss	11	R-R	5-9	168	12/9/1853	517	4543	35	308	191	.617	4500	1746	856	2.68
Rixey, Eppa	21	R-L	6-5½	210	5/3/1891	692	4494	39	266	251	.515	4633	1350	1082	3.15
Roberts, Robin	19	Bo-R	6-1	200	9/30/1926	676	4689	45	286	245	.539	4582	2357	902	3.40
Ruffing, Red	22	R-R	6-1½	215	5/3/1905	624	4342	46	273	225	.548	4294	1987	1541	3.80
Rusie, Amos	10	R-R	6-1	210	5/31/1871	412	3772	31	241	158	.604	3177	1953	1704	3.07
Ryan, Nolan	27	R-R	6-2	195	1/31/1947	807	5386	61	324	292	.526	3923	5714	2795	3.19
Seaver, Tom	20	R-R	6-1	206	11/17/1944	656	4782	61	311	205	.603	3971	3640	1390	2.86
Spahn, Warren	21	L-L	6-0	185	4/23/1921	750	5246	63	363	245	.597	4830	2583	1434	3.08
Sutton, Don	23	R-R	6-1	185	4/2/1945	774	5282	58	324	256	.559	4692	3574	1343	3.26
Vance, Dazzy	16	R-R	6-1	200	3/4/1891	442	2967	31	197	140	.585	2809	2045	840	3.24
Waddell, Rube	13	L-L	6-1½	196	10/13/1876	407	2958	50	191	142	.574	2480	2310	2310	2.16
Walsh, Ed	14	R-R	6-1	193	5/14/1881	432	2969	58	195	126	.607	2335	1731	617	1.82
Welch, Mickey	13	R-R	5-8	160	7/4/1859	566	4794	41	307	209	.595	4646	1841	1297	2.71
Wilhelm, Hoyt	21	R-R	6-0	190	7/26/1923	1070	2254	5	143	122	.540	1757	1610	778	2.52
Wynn, Early	23	Bo-R	6-0	235	1/6/1920	691	4566	49	300	244	.551	4291	2334	1775	3.54
Young, Cy	22	R-R	6-2	210	3/29/1867	906	7377	77	511	313	.620	7078	2819	1217	2.63

Most major-league teams publish yearbooks at the beginning of each season. The yearbooks include pictures of each player, their season and career records, some biographical information, and a summary of the previous season.

Now you can create your own team yearbook for your Little League team or for any team that you may be part of. Begin by taking pictures of each of your teammates. Remember to ask someone to take a picture of you! You can use a digital camera or a regular camera. Try to vary the pictures; that is, use different poses so your book will have some variety. Some pictures can be head-and-shoulder shots, and others can be action pictures, or posed action pictures. You can get good ideas from your favorite major-league team's yearbook or any baseball magazine.

Once you have all the pictures developed, you can plan the layout of your yearbook. As a suggestion, use 8 1/2 x 11 sheets. These are very easy to find in drug stores or school supply stores. Measure how much space you want to reserve for each picture or pictures on one page, making sure to allow enough space for whatever text you want to provide. You should write or type the text on the page before you paste on the pictures. Don't forget a page with all the names and ages of your teammates. You might even want to add their heights and weights, favorite major-league players, favorite teams, and so on. You can be as creative as you wish.

You'll want to have copies of the yearbook made for your teammates and others. Take your sheets to a copy shop, or ask someone to make photocopies for you. Then you can bind the yearbooks with a stapler, or punch three holes near the edge and place the yearbook in a three-hole portfolio. These can also be purchased at school supply stores or drugstores.

Think about asking a local businessperson to sponsor your yearbook. In turn, for helping you pay for the expenses of the yearbook, you could print a large advertisement of the business. Taking this one step further, think about placing lots of ads and distributing your yearbook to stores in your neighborhood or town. You might be able to raise enough money for new supplies for your team or for keeping your field maintained. There are many possibilities. Local businesses are often very eager to help with projects like this. Good luck!

Lost Seasons

It is fun to calculate what some players' final records might have been had they not lost playing time to injuries, or, as in the case of this project, to military service. Bob Feller and Ted Williams, two of baseball's most outstanding players, each lost three years in the prime of their careers to World War II. Williams lost most of two seasons to the Korean War, as well.

The career records for each player are reproduced on the student sheet, with questions for the students to answer. Some questions require library research for completion. Suggest to the students that they ignore the partial-season figures for Williams in 1952 and 1953, and make their calculations as if he did not play at all in those years. Do the same for Feller in 1945. The final questions, in which the students are asked to guess whether the players might have done better or worse during the lost seasons, are obviously open-ended. Some students might guess that the players would have done better than average during the World War II years because those years came in the midst of their most productive seasons.

Answers: (These revised records are based on dividing the career totals by the number of years played. This computes an "average" season's productivity. Add the totals for each lost season to the current lifetime totals for new totals. Remember to delete the 1952 and 1953 seasons for Williams and the 1945 season for Feller.)

Feller:
- Wins: 314 (tie for sixteenth on all-time list)
- Strikeouts: 3016 (thirteenth all-time)
  Winning Pct: .629

Williams:
- Home runs: 611 (fifth all-time)
- Hits: 3115 (nineteenth all-time)
- Lifetime BA: .344 (fifth all-time)

From *Baseball Math: Grandslam Activities and Projects for Grades 4–8* published by Good Year Books. Copyright © 2001, 2005 Christopher Jennison.

On this and the next page are the career records of two of baseball's greatest players, Ted Williams and Bob Feller. But their records would have been even greater if they had not lost several seasons to military service. Both players were in the service for three years during World War II, and Williams also lost most of two seasons during the Korean War.

Williams, Theodore (Ted) Samuel

Born, San Diego, California, August 30, 1918.
Bats Left. Throws Right. Height, 6'4". Weight, 195 pounds.

Year	Team	Lea.	Pos.	G	AB	R	H	2B	3B	HR	RBI	SB	Avg.
1939	Boston	A.L.	OF	149	565	131	185	44	11	31	145	2	.327
1940	Boston	A.L.	OF-P	144	561	134	193	43	14	23	113	4	.344
1941	Boston	A.L.	OF	143	456	135	185	33	3	37	120	2	.406
1942	Boston	A.L.	OF	150	522	141	186	34	5	36	137	3	.356
1943–44–45 (in Military Service)													
1946	Boston	A.L.	OF	150	514	142	176	37	8	38	123	0	.342
1947	Boston	A.L.	OF	156	528	125	181	40	9	32	114	0	.343
1948	Boston	A.L.	OF	137	509	124	188	44	3	25	127	4	.369
1949	Boston	A.L.	OF	155	566	150	194	39	3	43	159	1	.343
1950	Boston	A.L.	OF	89	334	82	106	24	1	28	97	3	.317
1951	Boston	A.L.	OF	148	531	109	169	28	4	30	126	1	.318
1952	Boston	A.L.	OF	6	10	2	4	0	1	1	3	0	.400
1953	Boston	A.L.	OF	37	91	17	37	6	0	13	34	0	.407
1954	Boston	A.L.	OF	117	386	93	133	23	1	29	89	0	.345
1955	Boston	A.L.	OF	98	320	77	114	21	3	28	83	2	.356
1956	Boston	A.L.	OF	136	400	71	138	28	2	24	82	0	.345
1957	Boston	A.L.	OF	132	420	96	163	28	1	38	87	0	.388
1958	Boston	A.L.	OF	129	411	81	135	23	2	26	85	1	.328
1959	Boston	A.L.	OF	103	272	32	69	15	0	10	43	0	.254
1960	Boston	A.L.	OF	113	310	56	98	15	0	29	72	1	.316
Major League Totals				2292	7706	1798	2654	525	71	521	1839	24	.344

If he had not been called to military service, how many games might Feller have won? _____

How many strikeouts would he have recorded? _____

What might have been his lifetime won-lost percentage? _____

If Williams had not lost so many years to the military, how many home runs might he have hit? _____

What would have been his total number of hits? _____

What might have been his lifetime batting average? _____

Lost Seasons (cont'd.)

Feller, Robert William Andrew
Born: Van Meter, Iowa, November 3, 1918.
Bats Right. Throws Right. Height, 6'. Weight, 185 pounds.

Year	Team	Lea.	G	IP	W	L	Pct.	SO	BB	H	ERA
1936	Cleveland	A.L.	14	62	5	3	.625	76	47	52	3.34
1937	Cleveland	A.L.	26	149	9	7	.563	150	106	116	3.38
1938	Cleveland	A.L.	39	278	17	11	.607	240	208	225	4.08
1939	Cleveland	A.L.	39	297	24	9	.727	246	142	227	2.85
1940	Cleveland	A.L.	43	320	27	11	.711	261	118	245	2.62
1941	Cleveland	A.L.	44	343	25	13	.658	260	194	284	3.15
1942–43–44 (in Military Service)											
1945	Cleveland	A.L.	9	72	5	3	.625	59	35	50	2.50
1946	Cleveland	A.L.	48	371	26	15	.634	348	153	277	2.18
1947	Cleveland	A.L.	42	299	20	11	.645	196	127	230	2.68
1948	Cleveland	A.L.	44	280	19	15	.559	164	116	255	3.57
1949	Cleveland	A.L.	36	211	15	14	.517	108	84	198	3.75
1950	Cleveland	A.L.	35	247	16	11	.593	119	103	230	3.43
1951	Cleveland	A.L.	33	250	22	8	.733	111	95	239	3.49
1952	Cleveland	A.L.	30	192	9	13	.409	81	83	219	4.73
1953	Cleveland	A.L.	25	176	10	7	.588	60	60	163	3.58
1954	Cleveland	A.L.	19	140	13	3	.813	59	39	127	3.09
1955	Cleveland	A.L.	25	83	4	4	.500	25	31	71	3.47
	Major League Totals		551	3770	266	158	.627	2563	1741	3208	3.23

Look at your new figures for Williams's homers, hits, and lifetime batting average. Where would he stand now among the all-time leaders in these categories?

Homers:_____

Hits: _____

BA: _____

(Your school or local librarian will help you find reference books to use to find these statistics.)

Look at Feller's new figures. Where would he stand among League Leaders now?

Wins: _____

Strikeouts: _____

Do you think the players might have done better or worse than average during the years they were in the service? Why?

_____

_____

From *Baseball Math: Grandslam Activities and Projects for Grades 4–8* published by Good Year Books. Copyright © 2001, 2005 Christopher Jennison.

Batting and pitching records are provided on pages 86–92. Lifetime records for home runs, hits, batting averages, and runs batted in are listed. Pitching records include wins, saves, strikeouts, and earned-run averages. Also listed are single-season records in the same categories. Some records are shown by era, because it is not always meaningful to compare today's game to the way baseball was played in earlier seasons.

As you've discovered, baseball interest thrives on numbers and statistics. Some youngsters find endless fascination in comparing and predicting performances based on average and cumulative statistics. For example, in the lifetime home run category there are eleven active players. Based on these players' average number of home runs per season, it's possible to estimate how many home runs they will hit by the time they retire. You would need to estimate how many more years a player would be likely to play and also allow for some decrease in a player's seasonal total as he gets closer to retirement.

In the pitching categories, Nolan Ryan looks as if he will remain the all-time king of strikeout pitchers. Can anyone catch him? Only Roger Clemens and Randy Johnson have more than 4,000 career strikeouts among present-day pitchers. Based on an average number of strikeouts per season so far, how many more seasons would Clemens and Johnson have to pitch to exceed Ryan's total? How old would they be? How would that compare with Ryan's age when he retired?

These are just a few examples of problems and research activities that can be based on the game's abundant statistics and records.

Lifetime Pitching Records (through 2004)
Wins
(inactive players 280 or more)

1	Cy Young	511
2	Walter Johnson	417
3	Christy Mathewson	374
4	"Pete" Alexander	373
5	Warren Spahn	363
6	Kid Nichols	361
	Pud Galvin	
7	Tim Keefe	344
8	Steve Carlton	329
9	Roger Clemens	328
10	John Clarkson	326
	Eddie Plank	
11	Don Sutton	324
	Nolan Ryan	
12	Phil Niekro	318
13	Gaylord Perry	314
14	Tom Seaver	311
	Mickey Welch	
15	Charles Radbourne	308
16	Greg Maddux	305
17	Lefty Grove	300
	Early Wynn	
18	Tommy John	288
19	Bert Blyleven	287
20	Robin Roberts	286
21	Tony Mullane	285
22	Ferguson Jenkins	284
23	Jim Katt	283
24	Red Ruffing	273
25	Burleigh Grimes	270
26	Jim Palmer	268
27	Bob Feller	266
	Eppa Rixey	
28	Jim McCormick	265
29	Gus Wehing	264
30	Tom Glavine	262

Baseball's Honor Roll (cont'd)

Lifetime Pitching Records (through 2004)

Strikeouts

1	Nolan Ryan	5714
2	Roger Clemens	4317
3	Randy Johnson	4161
4	Steve Carlton	4136
5	Bert Blyleven	3701
6	Tom Seaver	3640
7	Don Sutton	3574
8	Gaylord Perry	3534
9	Walter Johnson	3508
10	Phil Niekro	3342
11	Ferguson Jenkins	3192
12	Bob Gibson	3117
13	Greg Maddux	2916
14	Jim Bunning	2855
15	Mickey Lolich	2832
16	Cy Young	2798
17	Frank Tanana	2773
18	Curt Schilling	2745
19	David Cone	2668
20	Pedro Martinez	2653
21	Chuck Finley	2610
22	Warren Spahn	2583
23	Bob Feller	2581
24	Jerry Koosman	2556
25	Tim Keefe	2533
26	Christy Mathewson	2502
27	Don Drysdale	2486
28	Jack Morris	2478
29	Mark Langston	2464
30	Jim Kaat	2461
31	Sam McDowell	2453
32	Luis Tiant	2416
33	Dennis Eckersley	2401
34	John Smoltz	2398
35	Sandy Koufax	2396
36	Charlie Hough	2362
37	Robin Roberts	2357
38	Kevin Brown	2347
39	Early Wynn	2334
40	Rube Waddell	2316
41	Juan Marichal	2303

Saves

1	Lee Smith	478
2	John Franco	424
3	Trevor Hoffman	393
4	Dennis Eckersley	390
5	Jeff Reardon	367
6	Randy Myers	347
7	Rollie Fingers	341
8	Mariano Rivera	336
9	John Wetteland	330
10	Roberto Hernandez	320
11	Rich Aguilera	318
12	Troy Percival	316
13	Robb Nen	314
14	Tom Henke	311
15	Rich Gossage	310
16	Jeff Montgomery	304
17	Doug Jones	301
18	Bruce Sutter	300
19	Jose Mesa	292
20	Rod Beck	286
21	Todd Worrell	256
22	Dave Righetti	252
23	Billy Wagner	246
24	Dan Quisenberry	244
	Armando Benitez	
25	Sparky Lyle	238
26	Ugueth Urbina	227
	Hoyt Wilhelm	
27	Gene Garber	218
28	Greg Olson	217

Baseball's Honor Roll (cont'd)

From *Baseball Math: Grandslam Activities and Projects for Grades 4–8* published by Good Year Books. Copyright © 2001, 2005 Christopher Jennison.

Earned Run Average (by era)
1942–1960

1	Hoyt Wilhelm	2.52
2	Whitney Ford	2.74
3	Harry Brecheen	2.92
4	Mort Cooper	2.97
5	Max Lanier	3.01
6	Hal Newhouser	3.05
7	Tiny Bonham	3.06
8	Warren Spahn	3.08
9	Sal Maglie	3.15
10	Ed Lopat	3.21
11	Bob Lemon	3.23
	Dizzy Trout	
12	Stu Miller	3.24
13	Bob Feller	3.25
	Dutch Leonard	

Earned Run Average (by era)
1961–2004

1	Pedro Martinez	2.71
2	Sandy Koufax	2.76
3	John Franco	2.84
4	Andy Messersmith	2.86
	Jim Palmer	
	Tom Seaver	
5	Juan Marichal	2.89
6	Rollie Fingers	2.90
7	Bob Gibson	2.91
8	Dean Chance	2.92
9	Rich Gossage	2.93
10	Don Drysdale	2.95
	Greg Maddux	
11	Mel Stottlemyre	2.97
12	Randy Johnson	3.07

Single-Season Pitching Records
Wins (by era)
1942–1960

1	Hal Newhouser, 1944	29
2	Robin Roberts, 1952	28
3	Dizzy Trout, 1944	27
	Don Newcombe, 1956	
4	Hal Newhouser, 1946	26
	Bob Feller, 1946	
5	Dave Ferriss, 1946	25
	Mel Parnell, 1949	
6	Johnny Sain, 1948	24
	Bobby Shantz, 1952	
7	13 players tied	23

1961–2004

1	Denny McLain, 1968	31
2	Sandy Koufax, 1966	27
	Steve Carlton, 1972	
	Bob Welch, 1990	
3	Sandy Koufax, 1965	26
	Jaun Marichal, 1968	
4	7 players tied	25

Strikeouts (by era)
1942–1960

1	Bob Feller, 1946	348
2	Hal Newhouser, 1946	275
3	Herb Score, 1956	263
4	Don Drysdale, 1960	246
5	Herb Score, 1955	245
6	Don Drysdale, 1959	242
7	Sam Jones, 1958	225
8	Hal Newhouser, 1945	212
9	Bob Turley, 1955	210
10	Sam Jones, 1959	201
	Jim Bunning, 1959	
11	Robin Roberts, 1953	198
	Sam Jones, 1955	
12	Sandy koufax, 1960	197

Baseball's Honor Roll (cont'd)

Single-Season Pitching Records
Strikeouts (by era)
1961–2004

1	Nolan Ryan, 1973	383
2	Sandy Koufax, 1965	382
3	Randy Johnson, 2001	372
4	Nolan Ryan, 1974	367
5	Randy Johnson, 1999	364
6	Randy Johnson, 2000	347
7	Randy Johnson, 2002	344
8	Nolan Ryan, 1977	341
9	Nolan Ryan, 1972	329
	Randy Johnson, 1998	
10	Nolan Ryan, 1976	327
11	Sam McDowell, 1965	325
12	Curt Schilling, 1997	319
13	Sandy Koufax, 1966	317
14	Curt Schilling, 2002	316
15	J. R. Richard, 1979	313
	Pedro Martinez, 1999	
16	Steve Carlton, 1972	310
17	Mickey Lolich, 1971	308
	Randy Johnson, 1993	

Earned Run Average (by era)
1942–1960

1	Spud Chandler, 1943	1.64
2	Mort Cooper, 1942	1.78
3	Hal Newhouser, 1945	1.81
4	Max Lanier, 1943	1.90
5	Hal Newhouser, 1946	1.94
6	Bill Pierce, 1955	1.97
7	Whitney Ford, 1958	2.01
8	Al Benton, 1945	2.02
9	Allie Reynolds, 1952	2.06
10	Ted Lyons, 1942	2.10
	Howie Pollet, 1946	
	Spud Chandler, 1946	
	Warren Spahn, 1953	
11	Dizzy Trout, 1944	2.12
	Roger Wolff, 1945	2.12

Earned Run Average (by era)
1961–2004

1	Bob Gibson, 1968	1.12
2	Dwight Gooden, 1985	1.53
3	Greg Maddux, 1994	1.56
4	Luis Tiant, 1968	1.60
5	Greg Maddux, 1995	1.63
6	Dean Chance, 1964	1.65
7	Nolan Ryan, 1981	1.69
8	Sandy Koufax, 1966	1.73
9	Sandy Koufax, 1964	1.74
	Ron Guidry, 1978	
	Pedro Martinez, 2000	
10	Tom Seaver, 1971	1.76
11	Sam McDowell, 1968	1.81
12	Vida Blue, 1971	1.82
13	Phil Niekro, 1967	1.87
14	Joe Horlen, 1964	1.88
	Sandy Koufax, 1963	
15	Kevin Brown, 1996	1.89
16	Pedro Martinez, 1997	1.90

Saves

1	Bobby Thigpen, 1990	57
2	Eric Gagne, 2003	55
	John Smoltz, 2002	
3	Trevor Hoffman, 1998	53
	Randy Myers, 1993	
	Mariano Rivera, 2004	
4	Eric Gagne, 2002	52
5	Rod Beck, 1998	51
	Dennis Eckersley, 1992	
6	Mariano Rivera, 2001	50
7	Francisco Cordero, 2004	49
8	Rod Beck, 1998	48
	Dennis Eckersley, 1990	
	Jeff Shaw, 1998	
9	Lee Smith, 1991	47
	Armando Benitez, 2004	
	Jason Isringhausen, 2004	
	Lee Smith, 1991	
10	Dave Righetti, 1986	46
	Jose Mesa, 1995	
	Bryan Harvey, 1991	
	Tom Gordon, 1998	
	Lee Smith, 1993	
	Mike Williams, 2002	
11	Bruce Sutter, 1984	45

Dennis Eckersley, 1988
Jeff Montgomery, 1993
Duane Ward, 1993
Randy Myers, 1997
Mariano Rivera, 1999
Antonio Alfonseca, 2000
Eric Gagne, 2004
Eddie Guardado, 2002
Bryan Harvey, 1993
Jose Mesa, 2002
Robb Nen, 2001
Dan Quisenberry, 1983
Mariano Rivera, 1999
Kazahiro Sasaki, 2001
John Smoltz, 2003

Lifetime Batting Records (through 2004)
Home Runs

1	Hank Aaron	755
2	Babe Ruth	714
3	Barry Bonds	703
4	Willie Mays	660
5	Frank Robinson	586
6	Mark McGwire	583
7	Sammy Sosa	574
8	Harmon Killebrew	573
9	Reggie Jackson	563
10	Rafael Palmeiro	551
11	Mike Schmidt	548
12	Mickey Mantle	536
13	Jimmie Foxx	534
14	Ted Williams	521
	Willie McCovey	
15	Ernie Banks	512
	Ed Mathews	
16	Mel Ott	511
17	Eddie Murray	504
18	Ken Griffey, Jr.	501
19	Fred McGriff	493
20	Stan Musial	475
21	Dave Winfield	465
22	Jose Canseco	462
23	Carl Yastrzemski	452
24	Jeff Bagwell	446
25	Dave Kingman	442
26	Andre Dawson	438
27	Frank Thomas	436
28	Juan Gonzalez	434
29	Cal Ripken, Jr.	431
30	Billy Williams	426
31	Jim Thome	423
32	Gary Sheffield	415
33	Darrell Evans	414
34	Duke Snider	407
35	Andres Gallarraga	399
36	Dale Murphy	398
37	Joe Carter	396
38	Graig Nettles	390
	Manny Ramirez	

Batting Average

1	Ty Cobb	.366
2	Rogers Hornsby	.358
3	Joe Jackson	.356
4	Lefty O'Doul	.349
5	Ed Delehanty	.347
6	Tris Speaker	.345
7	Ted Williams	.344
	Billy Hamilton	
8	Dan Brouthers	.342
	Babe Ruth	
	Harry Heilman	
9	Pete Browning	.341
	Willie Keeler	
	Bill Terry	
10	George Sisler	.340
	Lou Gehrig	
11	Todd Helton	.339
12	Tony Gwynn	.338
	Jesse Burkett	
	Nap Lajoie	
13	Riggs Stephenson	.336
14	Al Simmons	.334
	John McGraw	
15	Paul Waner	.333
	Eddie Collins	
	Mike Donlin	
	Cap Anson	

Baseball's Honor Roll (cont'd)

Lifetime Batting Records (through 2004)

Batting Average (cont'd.)

16	Stan Musial	.331
	Sam Thompson	
17	Heinie Manush	.330
18	Wade Boggs	.328
	Rod Carew	.
19	Honus Wagner	.327
20	Tip O'Neil	.326
21	Joe DiMaggio	.325
	Jimmie Foxx	
	Earle Combs	
	Bob Fothergill	
	Babe Herman	
	Vladimir Guerrero	

Hits

1	Pete Rose	4256
2	Ty Cobb	4191
3	Hank Aaron	3771
4	Stan Musial	3630
5	Tris Speaker	3515
6	Carl Yastrzemski	3419
7	Cap Anson	3418
8	Honus Wagner	3415
9	Paul Molitor	3319
10	Eddie Collins	3313
11	Willie Mays	3283
12	Eddie Murray	3255
13	Nap Lajoie	3251
14	Cal Ripken Jr.	3184
15	George Brett	3154
16	Paul Waner	3152
17	Robin Yount	3142
18	Tony Gwynn	3141
19	Dave Winfield	3110
20	Rickey Henderson	3055
21	Rod Carew	3053
22	Lou Brock	3023
23	Wade Boggs	3010
24	Al Kaline	3007
25	Roberto Clemente	3000
26	Sam Rice	2987
27	Sam Crawford	2961
28	Frank Robinson	2943

29	Willie Keeler	2932
30	Jake Beckley	2930
	Rogers Hornsby	
31	Al Simmons	2927
32	Rafael Palmeiro	2922
33	Zack Wheat	2884
34	Frank Frisch	2880
35	Mel Ott	2876
36	Babe Ruth	2873
37	Jesse Burkett	2850
38	Brooks Robinson	2848
39	Charlie Gehringer	2839
40	George Sisler	2812

Runs Batted In

1	Hank Aaron	2297
2	Babe Ruth	2212
3	Cap Anson	2076
4	Lou Gehrig	1995
5	Ty Cobb	1961
6	Stan Musial	1951
7	Jimmie Foxx	1921
8	Eddie Murray	1917
9	Willie Mays	1903
10	Mel Ott	1860
11	Carl Yastrzemski	1844
12	Barry Bonds	1843
13	Ted Williams	1839
14	Dave Winfield	1833
15	Al Simmons	1827
16	Frank Robinson	1812
17	Rafael Palmeiro	1775
18	Honus Wagner	1773
19	Reggie Jackson	1702
20	Cal Ripken, Jr.	1695
21	Tony Perez	1652
22	Ernie Banks	1636
23	Harold Baines	1628
24	Goose Goslin	1609
25	Nap Lajoie	1599
26	George Brett	1595
	Mike Schmidt	
27	Andrew Dawson	1591
28	Rogers Hornsby	1584
29	Al Kaline	1583

From *Baseball Math: Grandslam Activities and Projects for Grades 4–8* published by Good Year Books. Copyright © 2001, 2005 Christopher Jennison.

30	Jake Beckley	1575
31	Tris Speaker	1559
32	Willie McCovey	1555
33	Fred McGriff	1550
34	Harry Heilmann	1541
35	Willie Stargell	1540
36	Joe DiMaggio	1537
37	Sammy Sosa	1530
38	Sam Crawford	1525
39	Jeff Bagwell	1510
40	Mickey Mantle	1509

Single-Season Batting Records
Home Runs

1	Barry Bonds, 2001	73
2	Mark McGwire, 1998	70
3	Sammy Sosa, 1998	66
4	Mark McGwire, 1999	65
5	Sammy Sosa, 2001	64
6	Sammy Sosa, 1999	63
7	Roger Maris, 1961	61
8	Babe Ruth, 1927	60
9	Babe Ruth, 1921	59
10	Jimmie Foxx, 1932	58
	Hank Greenberg, 1938	
	Mark McGwire, 1997	
11	Luis Gonzalez, 2001	57
	Alex Rodriguez, 2002	
12	Hack Wilson, 1930	56
	Ken Griffey, Jr., 1997	
	Ken Griffey, Jr., 1998	
13	Babe Ruth, 1920	54
	Babe Ruth, 1928	
	Ralph Kiner, 1949	
	Mickey Mantle, 1961	
14	Mickey Mantle, 1956	52
	Willie Mays, 1965	
	George Foster, 1977	
	Mark McGwire, 1996	
	Alex Rodriguez, 2001	
	Jim Thome, 2002	
15	Ralph Kiner, 1947	51
	Johnny Mize, 1947	
	Willie Mays, 1955	
	Cecil Fielder, 1990	

16	Jimmie Foxx, 1938	50
	Albert Belle, 1995	
	Brady Anderson, 1996	
	Greg Vaughn, 1998	
	Sammy Sosa, 2000	

Batting Average (by era)
1942–1960

1	Ted Williams, 1957	.388
2	Stan Musial, 1948	.376
3	Ted Williams, 1948	.369
4	Stan Musial, 1946	.365
	Mickey Mantle, 1957	.365
5	Harry Walker, 1947	.363
6	Dixie Walker, 1944	.357
	Stan Musial, 1943	.357
7	Ted Williams, 1942	.356
8	Phil Cavaretta, 1945	.355
	Lou Boudreau, 1948	.355
	Stan Musial, 1951	.355
	Hank Aaron, 1959	.355
9	Billy Goodman, 1950	.354
10	Harvey Kuenn, 1959	.353

Batting Average (by era)
1961–2004

1	Tony Gwynn, 1994	.394
2	George Brett, 1980	.390
3	Rod Carew, 1977	.388
4	Larry Walker, 1999	.379
5	Todd Helton, 2000	.372
	Ichiro Suzuki, 2004	
	Tony Gwynn, 1997	
6	Andres Gallarraga, 1993	.370
	Barry Bonds, 2002	
7	Wade Boggs, 1985	.368
	Tony Gwynn, 1995	
8	Wade Boggs, 1988	.366
9	Larry Walker, 1998	.363
	Joe Torre, 1971	
10	Wade Boggs, 1983	.361
	Norman Cash, 1961	
11	Albert Pujols, 2003	.359

Baseball's Honor Roll (cont'd)

Single Season Batting Records
Batting Average (by era)
1961–2004 (cont'd.)

12	Alex Rodriguez, 1996	.358
13	Roberto Clemente, 1967	.357

Hits

1	Ichiro Suzuki, 2004	262
2	George Sisler, 1920	257
3	Lefty O'Doul, 1929	254
	Bill Terry, 1930	
4	Al Simmons, 1925	253
5	Rogers Hornsby, 1922	250
	Chuck Klein, 1930	
6	Ty Cobb, 1911	248
7	George Sisler, 1922	246
8	Ichiro Suzuki, 2001	242
9	Heinie Manush, 1928	241
	Babe Herman, 1930	
10	Jesse Burkett, 1896	240
	Wade Boggs, 1985	
	Darrin Erstad, 2000	
11	Willie Keeler, 1897	239
	Rod Carew, 1977	
12	Ed Delehanty, 1899	238
	Don Mattingly, 1986	
13	Hugh Duffy, 1894	237
	Harry Heilmann, 1921	
	Paul Waner, 1927	
	Joe Medwick, 1937	
14	Jack Tobin, 1921	236
15	Rogers Hornsby, 1921	235
16	Lloyd Waner, 1929	234
	Kirby Puckett, 1988	

Runs Batted In

1	Hack Wilson, 1930	191
2	Lou Gehrig, 1931	184
3	Hank Greenberg, 1937	183
4	Lou Gehrig, 1927	175
	Jimmie Foxx, 1938	
5	Lou Gehrig, 1930	174
6	Babe Ruth, 1921	171
7	Chuck Klein, 1930	170
	Hank Greenberg, 1935	
8	Jimmie Foxx, 1932	169
9	Joe DiMaggio, 1937	167
10	Sam Thompson, 1887	166
11	Al Simmons, 1930	165
	Lou Gehrig, 1934	
	Manny Ramirez, 1999	
	Sam Thompson, 1895	
12	Babe Ruth, 1927	164
13	Babe Ruth, 1931	163
	Jimmie Foxx, 1933	
14	Hal Trosky, 1936	162
15	Sammy Sosa, 2001	160
16	Hack Wilson, 1929	159
	Lou Gehrig, 1937	
	Ted Williams, 1949	
	Vern Stephens, 1949	
17	Sammy Sosa, 1998	158
18	Al Simmons, 1929	157
	Juan Gonzalez, 1998	
19	Jimmie Foxx, 1930	156
20	Ken Williams, 1922	155
	Joe DiMaggio, 1948	
21	Babe Ruth, 1929	154
	Joe Medwick, 1937	

From *Baseball Math: Grandslam Activities and Projects for Grades 4–8* published by Good Year Books. Copyright © 2001, 2005 Christopher Jennison.

"Casey at the Bat"

More than one hundred years ago, a young man named Ernest Lawrence Thayer published a poem titled "Casey at the Bat" in the pages of the *San Francisco Examiner* (June 3, 1888). It has been a favorite piece of American folklore ever since. It is great fun to read or listen to, but it's even more fun to perform. One person can recite the poem for the enjoyment of all, or you can involve your whole class by assigning parts to a few, and asking the rest of the class to play the part of the crowd.

Someone must play the narrator, and three other students must play the part of Casey, the pitcher, and the umpire. If you have enough room, other students can play the parts of the various fielders and the base runners. The poem is reprinted on the following page with notes in the margin to help you dramatize the action.

For follow-up activities, some students could write newspaper reports of the game. Other people could interview Casey and report his comments and feelings. And the questions below could be assigned individually, or discussed at the end of the dramatization.

1. What was the final score of the game, and in what inning does the action take place?

2. How many spectators were in the stands?

3. If Casey had 188 hits and 515 at bats before his last appearance, what was his batting average after he struck out?

4. Why were the two hitters preceding Casey in the lineup portrayed so hopelessly? A hitter of Casey's stature would normally bat third or fourth in the lineup, and the batters just ahead of him would be among the team's best hitters. What sort of manager would put two poor hitters up ahead of Casey? Why? Could Casey have been a pinch-hitter? Is there any evidence of this? What did the poet mean when he called Flynn a lulu and Blake a cake? Did these terms have different meanings a hundred years ago?

Finally, you could ask a few students to do some financial calculations. Assume that the Mudville team's expenses that day were $100,000 for salaries, maintenance, advertising, etc. After the students have remembered how many spectators were at the game, ask them to calculate how much the spectators would have to be charged for tickets in order for the team to make a profit of 20%. To make the project a little more interesting, imagine that 30% of the seats were box seats, 60% were grandstand seats, and 10% were bleacher seats. Grandstand seats should be priced at twice the rate of bleacher seats and box seats at twice the rate of grandstand seats.

"Casey at the Bat" (cont'd.)

The outlook wasn't brilliant for the Mudville nine that
 day;
The score stood four to two with but one inning more
 to play.
And then when Cooney died at first, and Barrows did
 the same,
A sickly silence fell upon the patrons of the game.

A straggling few got up to go in deep despair. The
 rest
Clung to that hope which springs eternal in the
 human breast;
They thought, "If only Casey could get a whack at
 that—
We'd put up even money now with Casey at the bat."

But Flynn preceded Casey, as did also Jimmy Blake,
And the former was a lulu and the latter was a cake;
So upon that stricken multitude grim melancholy sat,
For there seemed but little chance of Casey's getting
 to the bat.

But Flynn let drive a single, to the wonderment of all,
And Blake, the much despised, tore the cover off the
 ball;
And when the dust had lifted, and the men saw what
 had occurred,
There was Jimmy safe at second and Flynn a-hugging
 third.

Narrator reads

1. **Pantomime two
 runners out at first.**
2. **Crowd sighs, then falls
 silent.**
3. **A few get up and leave
 in disgust.**

4. **The crowd sighs again.**

5. **Pantomime runners
 reaching second and
 third.**

From *Baseball Math: Grandslam Activities and Projects for Grades 4–8* published by Good Year Books. Copyright © 2001, 2005 Christopher Jennison.

From *Baseball Math: Grandslam Activities and Projects for Grades 4–8* published by Good Year Books. Copyright © 2001, 2005 Christopher Jennison.

Then from five thousand throats and more there rose a
    lusty yell;
It rumbled through the valley, it rattled in the dell;
It knocked upon the mountain and recoiled upon the
    flat,
For Casey, mighty Casey, was advancing to the bat.

6. Crowd cheers.

There was ease in Casey's manner as he stepped into
    his place;
There was pride in Casey's bearing and a smile on
    Casey's face.
And when, responding to the cheers, he lightly doffed
    his hat,
No stranger in the crowd could doubt 'twas Casey at
    the bat.

**7. Casey's gestures
should be very
extravagant.**

Ten thousand eyes were on him as he rubbed his
    hands with dirt;
Five thousand tongues applauded when he wiped
    them on his shirt.
Then while the writhing pitcher ground the ball into
    his hip,
Defiance gleamed in Casey's eye, a sneer curled
    Casey's lip.

8. More cheers.

And now the leather-covered sphere came hurtling
    through the air,
And Casey stood a-watching it in haughty grandeur
    there.
Close by the sturdy batsman the ball unheeded
    sped—
"That ain't my style," said Casey. "Strike one," the
    umpire said.

9. More pantomime.

**10. Casey and umpire
speak.**

From the benches, black with people, there went up a
    muffled roar,
Like the beating of the storm-waves on a stern and
    distant shore;
"Kill him! Kill the umpire!" shouted some one on the
    stand;
And it's likely they'd have killed him had not Casey
    raised his hand.

11. Crowd reacts.

With a smile of Christian charity great Casey's visage
    shone;
He stilled the rising tumult; he bade the game go on;
He signaled to the pitcher, and once more the
    spheroid flew;
But Casey still ignored it, and the umpire said, "Strike
    two."

12. More grand gestures from Casey.

"Fraud!" cried the maddened thousands, and echo
    answered "Fraud!"
But one scornful look from Casey and the audience
    was awed.
They saw his face grow stern and cold, they saw his
    muscles strain,
And they knew that Casey wouldn't let that ball go by
    again.

13. Crowd reacts.

The sneer is gone from Casey's lip, his teeth are
    clenched in hate;
He pounds with cruel violence his bat upon the plate.
And now the pitcher holds the ball, and now he lets it
    go,
And now the air is shattered by the force of Casey's
    blow.

14. More pantomime.

Oh, somewhere in this favored land the sun is shining
    bright;
The band is playing somewhere, and somewhere
    hearts are light.
And somewhere men are laughing, and somewhere
    children shout;
But there is no joy in Mudville—mighty Casey has
    struck out.

15. Background effects as available.

This project is a simplified version of the various forms of "Fantasy League" baseball that have become very popular in the past several years. It is recommended for only the most knowledgeable of your baseball fanatics. Basically a student "owner" selects twenty-two players from major-league rosters for his or her team, and after a predetermined period, the teams are ranked according to three batting and three pitching statistics.

Skills in resource allocation, cooperative learning, problem solving, and calculation are used throughout this project. The only materials needed are copies of your Sunday newspaper's sports section in which the statistics of all major-league players are updated. If your local paper does not include this information, you can get it from the Tuesday edition of *USA Today*, which includes American League statistics, and the Wednesday edition of *USA Today* for National League statistics. *Baseball Weekly*, a *USA Today* publication that comes out every Wednesday, includes totals for both leagues. Team "owners" can keep track of their teams' performances using the form provided on pages 99–100.

It is more realistic if student owners select their teams from the same division or league. But in order to make sure that there are enough players to go round, the number of student teams should be one less than the actual number of teams in a major league division or league. For instance, no more than four student owners should select from the American League East division team rosters, and no more than thirteen owners should select from all the American League rosters. The same would be true, of course, for the National League divisions and league.

Although the project can start at any time during the baseball season, and last as long as you and the students decide, it is suggested that you begin on opening day, in early April, and finish in early June, after an eight-week season. Beginning in September would only allow a one-month season.

Let's assume that you have selected four student-owners, and that they will select their twenty-two-man rosters from the American League East division. The first thing that occurs is the draft, and this should take place just a few days before the major league season begins. Owners will select their players from opening-day rosters that may appear in your Sunday newspaper just before opening day. If not, the official roster will appear in *USA Today* or *Baseball Weekly* issues that are published just prior to opening day. Each owner has a salary budget of $22.00. (Real money is not used.) Minimum salaries are ten cents, and the minimum bidding increment is ten cents. Make sure owners keep track of what they spend, and that they don't exceed their budgets.

A team consists of five outfielders, two catchers, one first baseman, one second baseman, one shortstop, one third baseman, one middle infielder (MI) (shortstop or second baseman), one cornerman (CM) (first or third base), one utility man (UM) (National League), one designated hitter (DH) (American League), and eight pitchers. Players must be eligible to play the positions for which they are selected. Eligibility is based on their appearing at that position in at least twenty games during the previous season. That information will appear in newspapers just prior to the start of the major-league season. Once the students' season has begun, player eligibility is based on the position played on opening day.

An owner begins the draft with a minimum bid for a player of ten cents. (It doesn't matter who starts the bidding.) The bidding then continues at ten-cent intervals until there is one bidder left. Owners need to be warned, of course, not to spend too much money on one player, or risk not having enough salary to fill out their rosters. Players not drafted go into the reserve pool, which is used later in the season to replace injured or traded players. The draft will take at least two to three hours, longer if more teams are involved, so it's best to schedule it for after school or the weekend before the season begins.

Once the major league season starts, the student-owners keep track of their players' performances using the form on pages 99–100. Box scores appear in daily newspapers and *Baseball Weekly*. Many Sunday sports sections provide cumulative statistics, as do *Baseball Weekly* and *USA Today*.

At the end of the first week, and once a week thereafter, the owners will need to hold an Exchange, during which an owner may release players (a) placed on a Major-League disabled list, (b) sent down to the minors, (c) traded to the other division or league, or (d) released by their "real" team. These players are replaced by players from the reserve pool who are eligible for the position they are replacing.

Injured players are placed in the reserve pool, as are players sent to the minors once they return to their major league teams. These players can be reclaimed during the next Exchange.

Transactions during Exchange take place in reverse order of the most recent team standings. Therefore, the last place team in the standings has the first option to replace a lost player. An owner can make only one transaction at a time, so if he or she needs to replace more than one player, the second transaction cannot take place until every other owner has had a chance to make a first transaction. Remind owners that players released into the reserve pool can be claimed by another team a week later. So, if a star player is injured, but predicted to be out of action for only a week or so, the owner might want to hold on to him, rather than give him up for a lesser player. The star could be reclaimed, but the previous owner might not have first option at that point.

There are three batting and three pitching statistics used to decide the winning team. Owners must keep track of these statistics daily so weekly team standings will be known on Exchange day. In a six-team league, the team with the most home runs gets six points, the second place teams gets five points, and so on. The same scoring method is used for the pitching records. In the unlikely event of a tie, the winner could be the owner with the most money left in his or her "account."

HINT: After reviewing the rules with your student-owners, you may want to name a "Commissioner" who would be responsible for enforcing the rules throughout the season.

From *Baseball Math: Grandslam Activities and Projects for Grades 4–8* published by Good Year Books. Copyright © 2001, 2005 Christopher Jennison.

Team Name: _____ Owner's Name _____

Pos.	Player		Week 1	Week 2	Week 3	Week 4	Week 5	Week 6	Week 7	Week 8
OF		Avg.								
		Hits								
		HRs								
OF		Avg.								
		Hits								
		HRs								
OF		Avg.								
		Hits								
		HRs								
OF		Avg.								
		Hits								
		HRs								
OF		Avg.								
		Hits								
		HRs								
C		Avg.								
		Hits								
		HRs								
C		Avg.								
		Hits								
		HRs								
1B		Avg.								
		Hits								
		HRs								
2B		Avg.								
		Hits								
		HRs								
3B		Avg.								
		Hits								
		HRs								
MI		Avg.								
		Hits								
		HRs								
SS		Avg.								
		Hits								
		HRs								
CM		Avg.								
		Hits								
		HRs								
DH		Avg.								
(or)		Hits								
UM		HRs								
TOTALS										

Leagues of Their Own (cont'd.)

Team Name: _____ **Owner's Name** _____

Pos.	Player		Week 1	Week 2	Week 3	Week 4	Week 5	Week 6	Week 7	Week 8
P		W								
		SVs								
		ERA								
P		W								
		SVs								
		ERA								
P		W								
		SVs								
		ERA								
P		W								
		SVs								
		ERA								
P		W								
		SVs								
		ERA								
P		W								
		SVs								
		ERA								
P		W								
		SVs								
		ERA								
P		W								
		SVs								
		ERA								
TOTALS										

From *Baseball Math: Grandslam Activities and Projects for Grades 4–8* published by Good Year Books. Copyright © 2001, 2005 Christopher Jennison.

Answers are provided for most activities.
For activities that do not appear here, answers vary.

Activities

2–Place Hitters
1. $20.00
2. 3
3. 70
4. $15.00
5. 9
6. 5
7. 3
8. 20

3–Card Profits
1. made $.20
2. made $.15
3. made $.35
4. $1.00

4–Circling the Bases
1. 360 feet
2. 240 feet
3. 240, 72.7%

5–Change Champ
1. $.50
2. $5.50
3. $2.50
4. $.25
5. $.75
6. $2.50
7. $3.00
8. $9.50
9. $2.75
10. $4.00

6–Team Trip
1. 1,073
2. 976
3. 582
4. 875
5. 1,180
6. 1,457

7. 878
8. Chicago and St. Louis
9. Boston and Houston
10. Boston—it travels the most miles
11. St. Louis

7–Diamond Data
1. 31 inches
2. 30 inches
3. Answers will vary.
4. 5'8"
5. 61 pounds
6. Answers will vary.

9–Sporting Goods
1. $4.89
2. $1.73
3. $12.49
4. $3.09
5. $.26
6. $2.95
7. $5.18
8. $1.36

10–A Trip to the Park
1. 30 minutes
2. 30 minutes
3. $4.10
4. $13.20
5. Answers will vary.
6. Answers will vary.
7, 8, and 9. Opinions should be supported.

11–A Game of Inches
1. Answers will vary in all cases.

12–Choosing Sides
1. Answers will vary, depending on selections.

13–Graphing Favorites
1. 6
2. 2
3. 9
4. Blue Jays
5. Rangers
6. Dodgers

14–Average Attendance
1. 8,093
2. 8,729
3. 6,298
4. 7,127
5. 8,609
6. 9,388
7. 7,044
8. 6,403
9. More, as 10,926 is higher than average for first three games

15–Crystal (Base) Ball
1. No-hitter pitched by a chimpanzee
2. Player steals eight bases in one inning.
3. Doubleheader is played on the moon.
4. Players agree to play for no money.

16–Big Bucks
1. 25 billion
2. 15 billion
3. decrease
4. 10 billion
5. opinion
6. 3.65 billion
7. 3.65 billion
8. 3.65 billion
9. 2.19 billion
10. 1.46 billion

From *Baseball Math: Grandslam Activities and Projects for Grades 4–8* published by Good Year Books. Copyright © 2001, 2005 Christopher Jennison.

Answer Key (cont'd.)

17–Hit Parade
1. July 2, July 2, July 5
2. July 2
3. July 6
4. July 10
5. 7
6. Cardinals
7. Tigers
8. Indians

18–Time Out
Answers will vary.

19–Baseball Budgets
1. 7%
2. six out of every hundred
3. transportation
4. taxes
5. administration
6. insurance
7. player salaries
8. one out of every five, or twenty out of every hundred

20–Growth Stocks
Estimates will vary.

21–Inning Time
Estimates will vary.

22–Pitching Pros
1. Martinez
2. 2
3. 2
4. 15
5. 15, 16, 20, 21, 23
6. 17

23–Where They Stand
1. 6
2. 7
3. 7
   The Cosmos will finish first, one game ahead of the Black Holes.

24–Slugfest
First: Hank
Second: Aaron
Third: Champ
Fourth: Homer
Fifth: Slugger

25–Game Goodies
Answers will depend on estimates and will vary.

26–Lining Up
1. 2
2. 12
3. 10
4. fifth
5. second
6. 44

27–Tips for Tips
1. $2.00, $14.25
2. $9.00, $72.38
3. $1.00, $8.80
4. $45.53
5. $22.95

28–Homer Daze
1. 2
2. Wednesday
3. Tuesday
4. Wednesday
5. 50
6. Answers will vary, but anything between two and six would be reasonable

29–Collecting Cards
1. Answers will vary.
2. 12, 24, 36, 48, 60, 72
3. 144
4. Approximately 89, depending on how student rounded off numbers

30–Wait 'til Next Year
1. Minnesota Twins
2. Twins' percentage in 2001 was .525; in 2002 it was .584—an increase of 11%
3. Detroit Tigers
4. .407
5. 16%

31–Home-run Showdown
1. 16%
2. 44
3. 15 at bats
4. 143
5. a little less than 1

32–Better Than the Boys?
1. 27
2. about 47%
3. about 41,000 at 48%

33–The Ballpark Factor
1. Red Sox
2. Yankees
3. Braves
4. Braves

34–Playing Favorites
Answers will vary.

From *Baseball Math: Grandslam Activities and Projects for Grades 4–8* published by Good Year Books. Copyright © 2001, 2005 Christopher Jennison.

From *Baseball Math: Grandslam Activities and Projects for Grades 4–8* published by Good Year Books. Copyright © 2001, 2005 Christopher Jennison.

35–Bigger Bucks

1. $17 million
2. $131,000
3. Answers to questions 3–6 will vary.
7. $3,750,000
8. at least $375,000 per season

36–Skill Sharpeners

1. 8 hours, 40 minutes
2. 1 hour, 30 minutes
3. 45 minutes
4. 6 hours, 40 minutes
5. 2 hours, 50 minutes
6. 14 hours, 30 minutes

37–Uniform Numbers

Kelsey: 7
Sonia: 11
Gary: 17
Lionel: 4
Chris: 100
Ellen: 25

38–Long Arms

Carrie: 26.4 m
Roy: 15.7 m
Kyle: 21.18 m
Ingrid: 26.34 m
Eduardo: 26.34 m

1. Roy
2. Ingrid
3. 13.28 m
4. 8.00 m
5. 36.46 m
6. Ingrid

39–Belting Beltran

1. $106,250
2. $168,317
3. Answers will vary.
4. $24,809
5. Answers will vary.
6. Answers will vary.

Challenge Problem

Answers will vary. For example, in 2004 Albert Pujols earned $7,100,000 and had 196 hits, 46 home runs, and 123 runs-batted-in. Based on his salary, he returned much better value to his team than Beltran.

40–Striking Out

1. Phil 1.73 to 1
2. Karen 1.13 to 1
3. .352
4. about 47
5. .294
6. about 59
7. Patty .22
8. Answers will vary.

41–Full House

1. 29
2. 60
3. 306
4. 105
5. 30
6. 20
7. 109
8. 70

42–Ballpark Field Trip

1. How many buses will be needed?
2. How many students will not be able to go to the game if only eleven buses show up?
3. How many seats are in the 23rd row?
4. The ballpark has 4,400 seats. Each row holds 200 seats. How many rows are there?
5. The ballpark has 4,600 seats. Each row holds 200 seats. How many rows are there? Note: Other answers are possible.

43–Finding the Average

1. 6,8,5,7
2. Ronnie: 3, Jenny: 2, Tom: 3, Bobbie: 3, Julio: 4
3. 8,530
4. Len: 9, Marian: 8, Sammy: 9, Chuck: 8, Stan: 5, Holly: 7, Michael: 6

44–Time for Baseball

Answers will vary.

45–Red Hots

1. Hot dog: $.60
2. Scorecard: about $.45
   Hats: $3.00
   Ball: $4.90
3. Soda pop: 200%
   Ice cream: $1.00

Answer Key (cont'd.)

46–Low Points

Butter Fingers: second
O. U. Stink: .066, fifth
Shakee Batts: 63, .080, fourth
Lowzee Hidder: 290, 6,889, .042, seventh
U. R. Badd: 4,117, 238, .058, sixth
Billy Goat: 328, 3,998, .082, third
Hesa Whiffer: 6,632, 101, .152, first
Bench Warmer: 22, 766, .029, eighth
Lefty Failure: second
Knuckles Flimsy: 77, 789, .098, third
Outa Control: 32/411, .078, fourth
Bull Pen: 66/972, .068, fifth
I. M. Crummy: 22,619, .036, sixth
Uneeda Curve: 71, 452, .157, first

Projects

54–Counting the Crowd

1. American League
2. Atlanta
3. N.Y. Mets
4. up

57–Player Targets

1. 135
2. 2,011
3. 6

From *Baseball Math: Grandslam Activities and Projects for Grades 4–8* published by Good Year Books. Copyright © 2001, 2005 Christopher Jennison.